GOLD AND SILVER.

AURIFEROUS GRAVELS IN THE WEAVERVILLE QUADRANGLE, CALIFORNIA.

By J. S. DILLER.

INTRODUCTION.

Few mining regions in California have attracted so much general attention and held it for so long a time as the Klamath Mountains of Siskiyou and Trinity counties, in the northwestern part of the State. The placers along Trinity and Klamath rivers were developed early in the gold rush and have been worked with varying energy to the present time. The La Grange mine, which is one of the largest hydraulic placers in the world, is now in the height of its activity.

In order that an account of mining in the Weaverville quadrangle may have its proper setting it is necessary to consider the general relations of the mountain ranges in the adjacent portions of California and Oregon.

THE KLAMATH MOUNTAINS.

GENERAL RELATIONS.

The mountain belt of the Pacific coast in California and Oregon includes a number of distinct ranges whose distribution and composition are in part illustrated by the accompanying geologic map (fig. 1). On the north are the Cascade Range and the Coast Range of Oregon, separated by the Willamette or Sound Valley as far south as Eugene. On the south are the Sierra Nevada and Coast Range of California, separated by the great valley of California. About the western part of the California-Oregon boundary, where all these ranges appear to meet, there is a distinct group of mountain ridges and peaks extending from a point north of the mouth of Rogue River in Oregon to Mad River in California. This group constitutes the Klamath Mountains. It embraces the South Fork, Trinity, and Salmon mountains of California and the Siskiyou and Rogue River mountains of Oregon.

These ranges are distinguished largely from geologic data, as will be more readily understood by referring to the map (fig. 1). The symbols on the map indicate in general the geologic age of the sedimentary rocks of the Klamath Mountains. To illustrate their areal distribution more clearly, all details and small areas have been omitted and outlines have been broadly generalized to cover large areas of igneous rocks. The map shows at a glance that the forma-

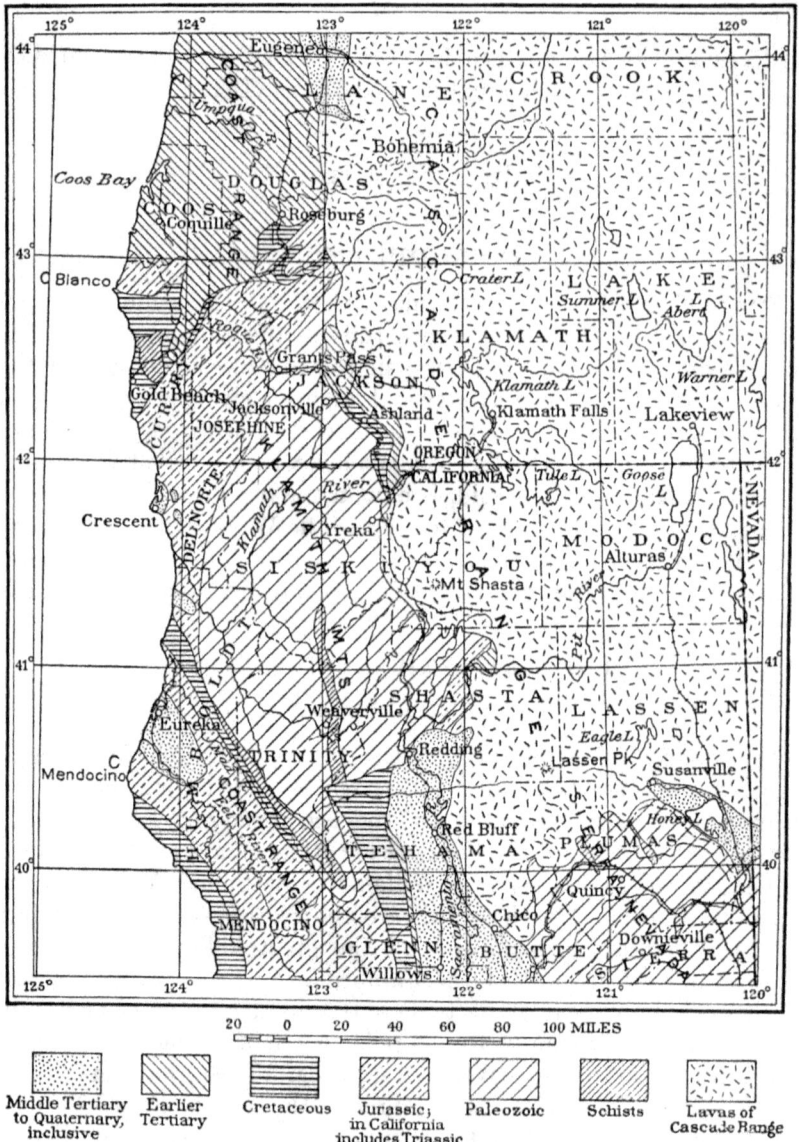

FIGURE 1.—Geologic map of the Klamath Mountains and adjacent ranges, California. Formations grouped and details generalized or omitted.

tions outlined occur mainly in the northwestern part of California, but they cover also a small area in Oregon.

The Klamath Mountains and the Sierra Nevada are composed in the main of essentially the same formations, and in the southern part of the Klamath Mountains the formations and lines of structure trend northwest and southeast, toward the Sierra Nevada, but in the northern portion they trend southwest and northeast, toward the Blue Mountains of eastern Oregon. Besides this general alignment of the formations of the Sierra Nevada, Klamath, and Blue mountains, their mineral resources are similar, a feature in which they are strongly contrasted with the Cascade and Coast ranges in Oregon and California.

GEOLOGIC HISTORY.

Little is known of the early geologic history of the Klamath Mountain region, yet it is evident that in pre-Devonian,[1] possibly in Algonkian [2] or late Archean [3] time the region was beneath the ocean, receiving the sediments from which the mica schist and intercalated crystalline limestones of South Fork Mountain and the Salmon Mountains north of Weaverville were formed.

The extensive development of Devonian and Carboniferous shales, sandstones, cherts, and limestones in the Klamath Mountain region shows that at least a part of the region continued beneath the sea through the whole or the greater part of the Paleozoic era, but the incompleteness of the succession and the discordance among the formations bear evidence of considerable earth movements at several times during the long period of sedimentation, culminating in the great mountain-building epoch at the close of the Jurassic. At times, too, while these sedimentary rocks were forming, especially before the Middle Devonian and during the later part of the Carboniferous and the greater portion of the Mesozoic, volcanoes were active in the region, giving rise to extensive sheets of contemporaneous lava and tuff intermingled with the sedimentary rocks and in many places covering them.

About the close of the Jurassic period this complex of sedimentary and igneous rocks was compressed, folded, faulted and uplifted to form the Klamath Mountains, and at the culmination of this process the mass was intruded by coarse granular bodies of plutonic rocks, such as granodiorite, gabbro, and peridotite, and by many dikes having a wide range in chemical and mineral composition.

As a consequence of this intense, varied, and long-continued igneous action, the heated circulating waters finally formed many ore deposits within the intruded masses or near their contacts. These deposits may have been enriched later by descending waters from the zone of oxidation.

Erosion and subsidence during the Cretaceous period reduced the Klamath Mountains to sea level, and for a brief interval they may

[1] Diller, J. S., Am. Jour. Sci., 4th ser., vol. 15, 1903, p. 343.
[2] Hershey, O. H., Am. Geologist, vol. 27, 1901, p. 245.
[3] Hershey, O. H., Am. Jour. Sci., 4th ser., vol. 30, 1912, p. 273.

have been completely covered by the ocean, for remnants of a once continuous sheet of conglomerates, sandstone, and shale are widely distributed in the region.

At the close of Cretaceous time the Klamath Mountains were again uplifted, and with a number of later oscillations and the consequent erosion they have been carved to their present form by streams, which have concentrated the gold in the auriferous gravels.

AURIFEROUS GRAVELS.

LOCATION AND EXTENT OF THE WEAVERVILLE QUADRANGLE.

The Weaverville quadrangle embraces portions of Shasta and Trinity counties, northwestern California, extending from Shasta post office west to the La Grange mine, and from Igo north nearly to Trinity Center. It is a little more than 34 miles in length and 26 miles in breadth, and its area is 905.27 square miles. It is bounded by parallels 40° 30' and 41° north latitude and meridians 122° 30' and 123° west longitude.

The Weaverville quadrangle is without railroads, but has good wagon roads connecting it with the main line of the Southern Pacific at Redding, Red Bluff, and Delta. Redding, at a distance of 52 miles from Weaverville, is the main distributing point.

FIELD WORK AND ACKNOWLEDGMENTS.

The principal area of auriferous gravels in the Weaverville quadrangle was roughly outlined on a small scale in 1893 [1] and on a larger scale in 1910.[2] The completion of the topographic map of the quadrangle in 1911 has made it possible to map these gravels in greater detail than heretofore. The work was only partly completed in the summer of 1912, and this is merely a brief preliminary report. Mr. H. G. Ferguson and I were associated in the field work and share equally the credit. The map (Pl. I) shows the areal results concerning the auriferous gravels thus far attained.

For information and assistance in the investigation I wish to express my hearty thanks to many mine owners and managers, but chiefly to Messrs. R. C. McDonald, of Trinity Center; J. C. Van Matre, and M. A. Singer, of Minersville; H. L. Lowden, of Weaverville; and Thomas McDonald, of French Gulch.

TOPOGRAPHY AND PHYSIOGRAPHY.

The Weaverville quadrangle has a mountainous topography. Like the rest of the Klamath Mountains, it is part of an uplifted peneplain [3] across which the larger streams have cut broad valleys [4]

---

[1] Fourteenth Ann. Rept. U. S. Geol. Survey, pt. 2, 1894, p. 414.
[2] Bull. U. S. Geol. Survey No. 470, 1911, p. 16.
[3] Bull. U. S. Geol. Survey No. 196, 1902, p. 15 (Klamath peneplain.)
[4] Idem, p. 49 (earlier valleys).

with even or slightly undulating bottoms so wide as to suggest secondary peneplains[1] and then intrenched themselves in deep canyons[2] with terraced slopes.

The relations of these important physiographic features, recording three cycles of erosion in the Klamath Mountains, are shown in a cross section of the Trinity River valley (fig. 2) a short distance south of Minersville, where the canyon of Trinity River, 1,500 feet in depth, trenches the Sherwood peneplain, well exposed on the long, even spurs of the west slope of Trinity Mountain and the gentle eastward slope of the flat table-land northwest of Buckeye Creek. In the Minersville region the broad divide between Stuart Fork and Rush Creek is called Buckeye Mountain, but as this local usage embraces under the one name features important to be distinguished, I refer herein to the table-land just mentioned as the Buckeye

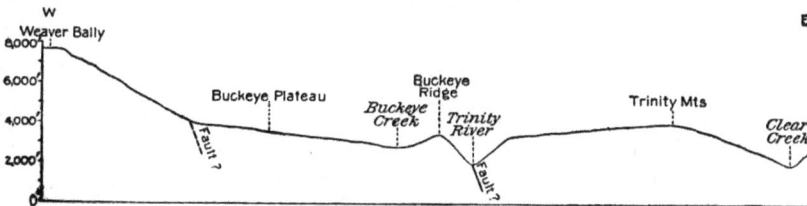

FIGURE 2.—Generalized section across Trinity River valley south of Minersville, Cal., showing the relation of Trinity River canyon to the Sherwood peneplain of Trinity Mountain and Buckeye Plateau and the Klamath peneplain of Weaver Bally.

Plateau and to the prominent ridge southeast of Buckeye Creek and adjoining Trinity River as the Buckeye Ridge.

Although remnants of the Klamath peneplain occur among the Salmon Mountains in the gentle slopes of the crests and peaks at altitudes of about 8,000 feet, they are not prominent in the Weaverville quadrangle. Elsewhere, however, as in the long, even crest of South Fork Mountain and at many other points in the Klamath Mountains, the Klamath peneplain is conspicuous though somewhat lower.

The complex history involved in the development of the physiographic features mentioned will not be discussed in detail in this preliminary report. It is sufficient to say here that all the auriferous gravel in the Klamath Mountains originated in the development of these features.

AREAL DISTRIBUTION OF THE GRAVELS.

On the accompanying map (Pl. I) is shown the areal distribution of the gravels, by solid boundary lines as far as the survey is completed, but by dotted lines where the survey is incomplete.

---

[1] Bull. U. S. Geol. Survey No. 196, 1902, p. 22 (Sherwood peneplain).

[2] Idem, p. 58 (later valleys); Bull. Dept. Geology Univ. California, vol. 3, No. 22, 1904, p. 425 (Sierran canyons).

The gravels already mapped may be considered as belonging to three areas. The oldest, largest, and most important is that of the Weaverville basin, which is associated with the Sherwood peneplain. This area extends from the La Grange mine to the East Fork of Stuart Fork and may have formerly extended to Trinity Center. Although not now a basin it was at one time and drained directly southwestward to Weaverville and the La Grange mine. The survey of this basin is incomplete. The second area is on Trinity River, extending from Trinity Center to Lewiston. These two areas are in Trinity County and are not subject to restrictions concerning débris. The third area is in Shasta County on Clear Creek, a tributary of the Sacramento, and in it the débris must be controlled.

In the two areas last mentioned the gravels are associated more or less closely in origin with the formation of the stream canyons, and are very much younger than most of the gravels of the Weaverville basin.

WEAVERVILLE BASIN.

The Weaverville basin area is irregularly triangular, being 24 miles in length if it extends to Trinity Center and 8 miles in greatest width. The base of the triangular area is near Weaverville and the apex at Trinity Center, but there is a wide interruption in the area due to recent glacial action along Swift Creek and East Fork of Stuart Fork. Southeast of Weaverville, in the vicinity of Browns Mountain and Lowden's ranch, there are a number of detached areas which have been included in considering the general form of the mass. Between East Fork of Stuart Fork and Weaverville, a distance of nearly 17 miles, the deposit is continuous, and has a width of 1 to 3 miles.

This large continuous body is not homogeneous. It varies much from place to place and consists of two more or less contrasted classes of material—one including many angular bowlders intermingled with angular to subangular, locally rounded fragments of sand and clay, as in glacial till; the other including for the most part well-rounded fragments, a few bowlders, and gravel commingled with a smaller proportion of sand and clay, as may occur in fluviatile deposits. The relations of these materials can not be discussed to advantage until the survey of the Weaverville basin is completed.

Gravels of the same age as much of the gravel in the Weaverville basin are now worked in the Red Hill mine, 1 mile west of Trinity Center, at an elevation of about 2,800 feet above the sea, or 500 feet above the town. These gravels are very much older than those extensively mined years ago along the western edge of Trinity Center and have been affected by profound earth movements which the later gravels have not experienced.

The gravel of the Red Hill mine is 100 to 125 feet in thickness. Besides being partly cemented it is tilted and faulted down into the

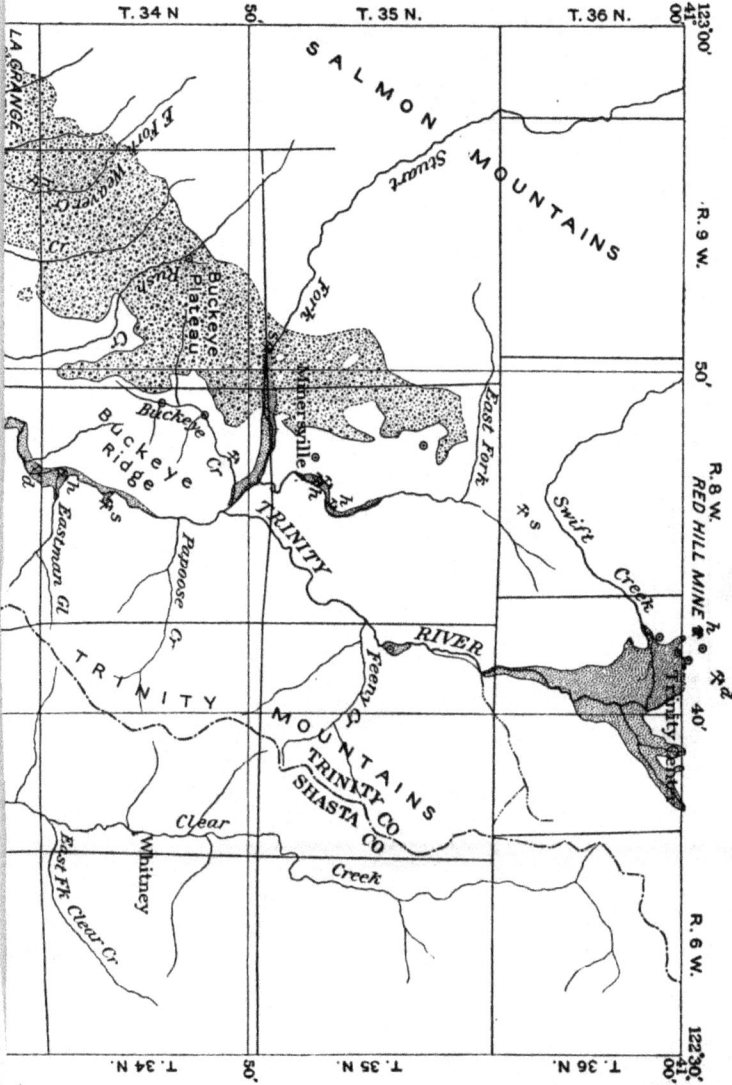

MAP SHOWING AURIFEROUS GRAVELS OF THE MIDDLE AND NORTHERN PORTIONS OF THE WEAVERVILLE QUADRANGLE, CALIFORNIA.

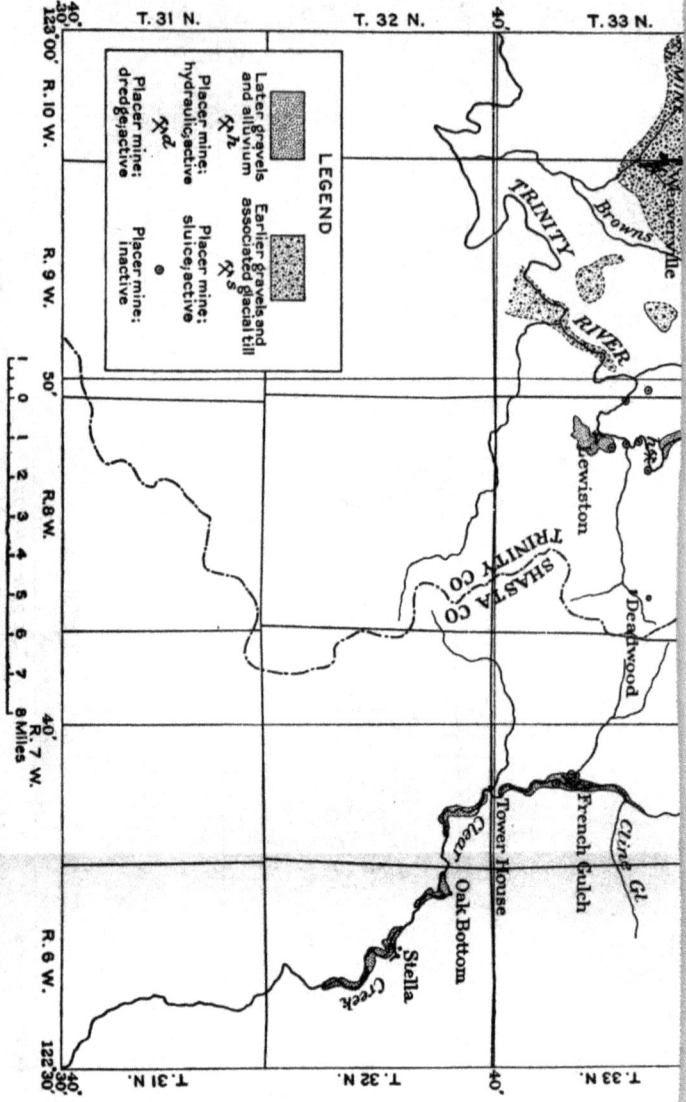

LEGEND

Later gravels and alluvium

Earlier gravels and associated glacial till

Placer mine; hydraulic; active — Xph

Placer mine; sluice; active — Xs

Placer mine; dredge; active — Xd

Placer mine; inactive

Scale: 1 0 1 2 3 4 5 6 7 8 Miles

T. 31 N.
T. 32 N.
T. 33 N.

R. 10 W.
R. 9 W.
R. 8 W.
R. 7 W.
R. 6 W.

40°30'
123°00'
50'
40'
40'
122°30'

TRINITY RIVER
Browns
Weaverville
Lewiston
Deadwood
French Gulch
Tower House
Oak Bottom
Stella
Clear Creek
Cline Gl.
Clear Creek

SHASTA CO
TRINITY CO

slate bedrock, which belongs to the Bragdon formation. Although 50 per cent of the gravel is less than 2 inches in diameter and the greater part is well rounded, there are, especially near the bottom, a number of bowlders from 3 to 10 feet in diameter. Fine whitish sand, containing large fragments of wood, occurs in places upon the bed-rock. Much of the gravel, especially in the upper portion, is com-pletely decomposed and colored pink. The soft pebbles cut like cheese. The lower portion contains pebbles and bowlders of con-glomerate from the Bragdon formation, and these are generally hard and smooth.

The gold is moderately coarse, and irregular nuggets worth as much as $30 are sufficiently smoothed by attrition to indicate transporta-tion to a considerable distance.

An area of this gravel at least several acres in extent is still available for mining on the divide between Brush and Hatchet creeks. That on the north slope of Hatchet Creek was mined some years ago. Much of it has been removed by the streams in eroding the present valleys, the gold thus being concentrated and the later gravels at lower levels in the same region enriched.

Between Swift Creek and East Fork small placers have been worked near Davis and Hubbard creeks, branches of the East Fork, about 3 miles north of Bowerman's ranch, but they show very little gravel. Their gold is in residual meta-andesite (greenstone), not far from its contact with the slates of the Bragdon formation.

Northwest of Minersville, between East Fork and Stuart Fork, in the region drained by Strope, Digger, and Mule creeks, a great body of fragmental material forms some of the prominent flat-topped divides that belong to the Sherwood peneplain. The material is bowldery and somewhat angular, resembling glacial till, and it may be at least in part of glacial origin. The bulk of it near the surface, and especially near the peneplain level, is completely decomposed, and in some places the decomposition products are highly colored. In the region northwest of Minersville the material resembling glacial till is not covered with gravel.

The general absence or scarcity of gold in this supposed glacial till, as shown by D. F. MacDonald,[1] has led many of the miners to call it "dead wash." The streams cut through the dead wash in places and derive gold from the underlying bedrock, which in that region is generally an altered ancient lava, meta-andesite.

The most successful placer mines in the Minersville region have been limited to the later gravels of the stream terraces about 100 feet above East Fork, within 4 miles north of Minersville. These bench gravels, as shown on the map, are entirely distinct from the so-called glacial till or dead wash of the hills farther west. The bedrock is

---

[1] Bull. U. S. Geol. Survey No. 430, 1910, p. 56.

slate (Bragdon formation) except in part of the Unity mine, near the west line of sec. 27, where the bedrock is meta-andesite. The gravel is generally well washed, contains some bowlders, and has a thickness of 40 to 100 feet, forming a terrace much of which yet remains to be mined. The water for the mine, which is now said to be under the control of the Trinity Gold & Hydraulic Dredging Co., is to be supplied from East Fork by a 12-mile ditch carrying 3,000 miner's inches to a 300-foot head.

Between Stuart Fork and Rush Creek there is a great body of fragmental deposits made up in part of more or less angular material that suggests glacial origin but including much gravel that is well rounded by water. The main belt crossing the Buckeye Plateau lies west of Buckeye Creek. It has a width apparently of nearly 3 miles and a depth where greatest of more than 800 feet. Three shafts from 40 to 195 feet in depth have been sunk on the top portion of this deposit to test it for gold. Little if any gold is said to have been found and no actual mining tests resulted.

The surface of the Buckeye Plateau is generally reddish soil to a depth of 15 feet or more, passing downward into a sandy argillaceous mass in which the forms of rounded to subangular fragments and some bowlders may be seen and a few well-rounded solid pebbles are preserved. On the plateau surface there are for the most part only a few scattered pebbles with here and there a small well-rounded bowlder, but on the slopes of the ravines cut by branches of Buckeye Creek, as well as on the plateau borders facing Rush Creek and Stuarts Fork, well-rounded gravel is in many places abundant and extensive. Buckeye Creek drains the great body of fragmental material that forms the Buckeye Plateau. The main portion of the creek was rich in placer gold, but it received its gold chiefly from the east side, where Dutch and Whitney gulches, draining contacts of slate and porphyry, were very much richer than the gulches on the west, heading in the plateau.

Rush Creek, whose present stream bed affords some good placers, cuts a deep, narrow valley directly across the old channel but does not reach the bottom.

The greater portion of the gravels of Weaverville basin lies southwest of Rush Creek, forming the divide between Rush and Browns creeks on one side and the branches of Weaver Creek on the other. In the vicinity of Weaverville the older gravels were in large measure worked over by modern streams and the gold was concentrated in the later gravels, which are rich and have been mined for many years.[1] The gold of the later gravels is derived chiefly from the older gravels, but the older gravels are in few places, if anywhere, so rich as to afford profitable placer ground. This fact explains why the immediate

---

[1] The inactive mines in that region are so numerous that no attempt has been made to map them.

vicinity of Weaverville has been almost completely washed in placers, while the gravel divides are as yet untouched.

The lower portion of the deposits in the Weaverville basin is generally much finer and more firmly cemented than the upper portion. In the mine formerly known as the Hupp mine it is conveniently used as bedrock for working the upper portion as a placer.

The La Grange mine and that of the Trinity River Consolidated Hydraulic Mining Co. (formerly the Hupp mine) are the most important placers in the Weaverville basin. Both of them have been noticed in earlier publications [1] of the Survey, and need only be referred to in this report.

TRINITY RIVER AREA.

The Trinity River area, so far as already mapped, includes the gravels from Trinity Center to Lewiston. All the gravel of this area belongs to the later portion of the canyon-cutting epoch. The gravels of this epoch may be conveniently referred to as later gravels, in comparison with the earlier gravels of the Weaverville basin area. The Red Hill mine, at the edge of this area, belongs geologically in the Weaverville basin and has been described on pages 10–11.

In the vicinity of Trinity Center, which has been a great placer camp for many years, the stream beds are deeply aggraded. On one side the gravels merge into the glacial material of Swift Creek and on the other they are covered by the alluvium of the present streams.

On the western border of Trinity Center an old river terrace, once nearly half a mile in length, has been mined away, leaving a gravel bluff in places over 100 feet in height. Above this bluff there are remnants of gravel at intervals on the slope leading up to the Red Hill mine, in which occur the earlier gravels belonging to the same formation as those of the Weaverville basin area, described on page 10.

A short distance north of Trinity Center, on a broad gravel flat near Trinity River, a bucket dredge has been successfully operated more or less continuously since 1903. Dredging was suspended in the summer of 1912 to install a chain of 61 new 7½-foot buckets and otherwise improve the machinery. Considerable ground has already been covered, and there is enough ahead to keep the dredge at work for several years. The success of this project has led to prospecting at other points in the vicinity, especially in the wide alluvial expanse within several miles south of Trinity Center, but as yet, so far as I am aware, no new dredging operations have been commenced.

Along Trinity River for more than 20 miles south of Trinity Center there has been very little placer mining, but at Mooney Gulch begins a region of greater activity. There is a placer mine on Mooney

[1] Bull. U. S. Geol. Survey No. 430, 1910, pp. 51–56; No. 470, 1911, pp. 16–18.

Gulch a quarter of a mile above its mouth and a hydraulic mine on the river at the mouth of Eastman Gulch. Both have been productive for some years.

Recently two new projects have been opened in that region with unusually extensive preparation. These are the dredge of the Trinity River Dredging Co. and the hydraulic mine of the Trinity River Mining Co.

The Trinity River Dredging Co. has erected an electric dredge on a gravel flat of Trinity River at the mouth of Jennings Gulch, about 4 miles above Lewiston. The dredge is said to have a chain of forty-four 11-foot buckets that can reach a depth of 40 feet. The power-house is 5 miles farther up Trinity River, near the mouth of Stuart Fork. The water is brought from Stuart Fork by a ditch 7½ miles in length and delivered at the powerhouse with a head of 285 feet. In September, 1912, with the power plant already running and the dredge nearly completed, the company was about to begin operations. There is a large amount of available ground along Trinity River below this point, and many of the people of that region are greatly interested in the success of this dredge.

A few years ago the Trinity River Mining Co. attempted to drain the bed of Trinity River at Big Bend, a mile north of Lewiston. A tunnel 1,385 feet in length was constructed, giving a head of 25 feet, which was utilized in a turbine and centrifugal pump to force water directly into pipes for hydraulic mining. The tunnel did not completely drain the river, but the available water power was used for hydraulicking some of the gravels in the vicinity. This property has been leased by the Horseshoe Placer Mining Co., which proposes to construct a concrete dam at the intake of the tunnel to complete the drainage of nearly a mile of the river bed.

The gravel of Trinity River for some distance above Lewiston has been aggraded in places to a depth of 30 feet and affords an opportunity for successful mining beneath the river. Near the mouth of Deadwood Gulch two shafts were sunk by the side of the river to a depth of 30 feet, passing through 17 feet of cement underlain by 8 feet of gravel. The gravel was mined out beneath the river for a distance of 267 feet.

CLEAR CREEK AREA.

The bed of Clear Creek from French Gulch to and beyond Stella, in Shasta County, was washed in many places years ago. At several points, especially in the vicinity of French Gulch, the mining included the terraces up to 100 feet. In like manner the beds of Whiskey Creek and other tributaries of Clear Creek were for the most part mined long since, so that comparatively little ground is available at the present time. Only two placer mines were reported in Shasta

County in 1911.   Both were in the vicinity of French Gulch and used ground sluicing.

On Clear Creek, as on Trinity River, much of the gold now found in the gravels is derived from the residual material at the contacts between the slates (Bragdon formation) and the more ancient lavas, or the porphyry dikes.   In many places these contacts may be well worth washing as placers.

OUTLOOK FOR PLACER MINING IN THE DISTRICT.

The outlook for future placer mining in this region is encouraging. The success of the La Grange mine consists in the economical treatment on a large scale of relatively low grade gravel.   The great body of fragmental material to which that of the La Grange mine belongs extends northeastward from the La Grange mine to and beyond Stuart Fork, and may include other bodies of gravel similar to that of the La Grange mine.   It lies parallel to and beneath the great ditch that supplies water to the La Grange mine, and any other masses of gravel in the same belt could therefore be easily tested on an appropriate scale.   The streams within the Weaverville basin, though partly aggraded, would generally afford a fair dumping ground for the higher gravel, and where the gravel is not too firmly cemented it might, with facilities equal to those of the La Grange, be economically mined.

The success of the dredging at Trinity Center and of the placers north of Lewiston gives confidence to those who are attempting larger developments at the mouth of Eastman Gulch and at the bend of Trinity River above Lewiston, and the region may well be regarded as worthy of investigation by capitalists interested in dredging and hydraulic mining.

GOLD LODES OF THE WEAVERVILLE QUADRANGLE, CALIFORNIA.

By HENRY G. FERGUSON.

SITUATION AND AREA.

The Weaverville quadrangle includes parts of Trinity and Shasta counties, Cal., and at its northwest corner touches the southern point of Siskiyou County. It lies between 40° 30′ and 41° north latitude and 122° 30′ and 123° west longitude, and covers an area of 905.27 square miles. The country is mountainous and sparsely populated. A few small mining camps such as Minersville, Trinity Center, French Gulch, and Whiskeytown were prosperous towns in the early days of the rich placer workings. Along the valley of Trinity River there is a small amount of farming land. The principal town is Weaverville, the county seat of Trinity County, with a population of 913. The city of Redding, on the Southern Pacific Railroad in Shasta County, serves as a distributing point for the entire quadrangle, except the extreme northeast corner. Daily stages run from Weaverville to Redding, a distance of 54 miles, and from Trinity Center to Delta.

The extreme southern and western portions of the quadrangle, including the Igo district, in Shasta County, the Bully Choop and Eastman Gulch districts, in Trinity County, and the Keswick (Iron Mountain) copper district already described by Graton,[1] are not covered by this report.

FIELD WORK AND PREVIOUS DESCRIPTIONS.

The writer had the opportunity of visiting the principal gold mines in the course of the areal geologic mapping of the quadrangle during the field season of 1912 as assistant to J. S. Diller. He is greatly indebted to Mr. Diller for criticism and suggestions and to all the mining men of the region for courtesy and hospitality.

---

[1] Graton, L. C., The occurrence of copper in Shasta County, Cal.: Bull. U. S. Geol. Survey No. 430, 1910, p. 71.

Very little has been written bearing directly upon the lode deposits of the region. The geology, physiography, and placer deposits have been discussed in papers by Diller,[1] Hershey,[2] and MacDonald.[3]

The area immediately to the east has been mapped and described by Diller [4] and the copper deposits of the Shasta copper belt by Graton.[5] The lode deposits of certain districts north of the Weaverville quadrangle have been described by Hershey [6] and MacDonald.[7] Hershey [8] has also discussed the origin of the pocket deposits which form a feature of part of this area.

GEOGRAPHY.

The Weaverville quadrangle lies entirely within the topographic province of the Klamath Mountains. Two streams, Clear Creek and Trinity River, with their tributaries, drain almost the whole area. Clear Creek rises in the extreme northeast corner and flows in a general southerly direction, leaving the quadrangle near Igo, in the southeast corner. This point is the lowest in the area, having an elevation of less than 800 feet above sea level. Clear Creek empties into Sacramento River 5 miles south of Redding. Trinity River enters the quadrangle from the north at Trinity Center and flows in a general southwesterly direction. It leaves the area at a point 5 miles southwest of Weaverville and flows northwestward for 60 miles to Klamath River. The divide between these two streams is the boundary between Trinity and Shasta counties. Two groups of mountains are conspicuous features of the topography. The Salmon Mountains, cut in two by Stuart Fork, reach a maximum elevation of 8,879 feet in an unnamed peak in the extreme northwest corner of the area. Along the southern border of the quadrangle are the

---

[1] Diller, J. S., Tertiary revolution in the topography of the Pacific coast: Fourteenth Ann. Rept. U. S. Geol. Survey, pt. 2, 1894, pp. 397–434; Revolution in the topography of the Pacific coast since the auriferous-gravel period: Jour. Geology, vol. 2, 1894, p. 32; Topographic development of the Klamath Mountains: Bull. U. S. Geol. Survey No. 196, 1902; Klamath Mountain section, California: Am. Jour. Sci., 4th ser., vol. 15, 1903, pp. 342–362; The Bragdon formation: idem, vol. 19, 1905, pp. 379–387; The auriferous gravels of the Trinity River basin, California: Bull. U. S. Geol. Survey No. 470, 1911, pp. 11–29. See also Auriferous gravels in the Weaverville quadrangle, in this volume, pp. 5–15.

[2] Hershey, O. H., Metamorphic formations of northwestern California: Am. Geologist, vol. 27, 1901, pp. 225–245; Some evidence of two glacial stages in the Klamath Mountains in California: idem, vol. 31, 1903, pp. 139–156; Structure of the southern portion of the Klamath Mountains, California: idem, pp. 231–245; Sierran valleys of the Klamath region: Jour. Geology, vol. 11, 1903, pp. 155–165; The Bragdon formation in northwestern California: Am. Geologist, vol. 33, 1904, pp. 248–256; The river terraces of the Orleans basin, California: Bull. Dept. Geology Univ. California, vol. 3, 1904, pp. 423–475.

[3] MacDonald, D. F., The Weaverville-Trinity Center gold gravels, Trinity County, Cal.: Bull. U. S. Geol. Survey No. 430, 1910, p. 50.

[4] Diller, J. S., Redding folio (No. 138), Geol. Atlas U. S., U. S. Geol. Survey, 1906.

[5] Graton, L. C., The occurrence of copper in Shasta County, Cal.: Bull. U. S. Geol. Survey No. 430, 1910, p. 71.

[6] Hershey, O. H., Gold-bearing lodes in California: Am. Geologist, vol. 25, 1900, pp. 76–96.

[7] MacDonald, D. F., Notes on the gold lodes of the Carrville district, Trinity County, Cal.: Bull. U. S. Geol. Survey No. 530, 1913, pp. 9–41.

[8] Hershey, O. H., Origin and age of certain gold deposits in northern California: Am. Geologist, vol. 24, 1899, pp. 38–43; Origin of gold pockets in northern California: Min. and Sci. Press, vol. 101, 1910, pp. 741–742.

Bully Choop Mountains, whose highest point, Bully Choop, has an elevation of 6,964 feet.

The greater part of the area consists of irregular sprawling ridges, brush covered or timbered, whose summits are remnants of a deeply dissected peneplain and which attain a maximum elevation of slightly over 5,000 feet along the northern border and grade down to about 3,500 feet near Whiskeytown. Along the eastern border of the quadrangle this upland falls off sharply into the north end of the broad Sacramento Valley.

GEOLOGY.

BROAD FEATURES.

The geologic history of the Klamath Mountains, of which the Weaverville quadrangle is a part, is summarized in the paper by J. S. Diller on pages 13–14, and hence a description of the rock formations will be sufficient for the purposes of the present paper. In the western part of the area are found older biotite and hornblende schists, the former containing lenses of crystalline limestone. In the eastern and central parts the Copley meta-andesite (Devonian or older)[1] is overlain unconformably by the Bragdon formation (Mississippian). Later movements which have affected these have left their mark in the irregular line of contact of these two formations and in the contortion of the Bragdon formation. In late Jurassic or early Cretaceous time[2] came a period of igneous activity producing the complicated series of granitic and porphyritic rocks which to-day occupy a large proportion of the surface. During Cretaceous and Tertiary time the region passed through several stages of erosion and base-leveling.

The accompanying geologic map (Pl. II) was made during the field season of 1912 by Mr. Diller and the writer and shows the principal rock formations of the area covered by the season's field work and the location of the more important lode mines.

SEDIMENTARY ROCKS.

The oldest sedimentary rock of the quadrangle is a biotite schist, a portion of the Salmon schist of Hershey,[3] a small mass of which outcrops between the serpentine and quartz diorite at the head of the East Fork of Trinity River. The schist contains some lenses of limestone that have been completely recrystallized by the intrusion of the quartz diorite, with the development of much garnet and epidote.

Except for the schist the pre-Carboniferous rocks of the quadrangle are represented only by the hornblende schist and meta-

---

[1] Diller, J. S., Redding folio (No. 138), Geol. Atlas U. S., U. S. Geol. Survey, 1906, p. 6.
[2] Idem, p. 10.
[3] Hershey, O. H., Am. Geologist, vol. 27, 1901, p. 225.

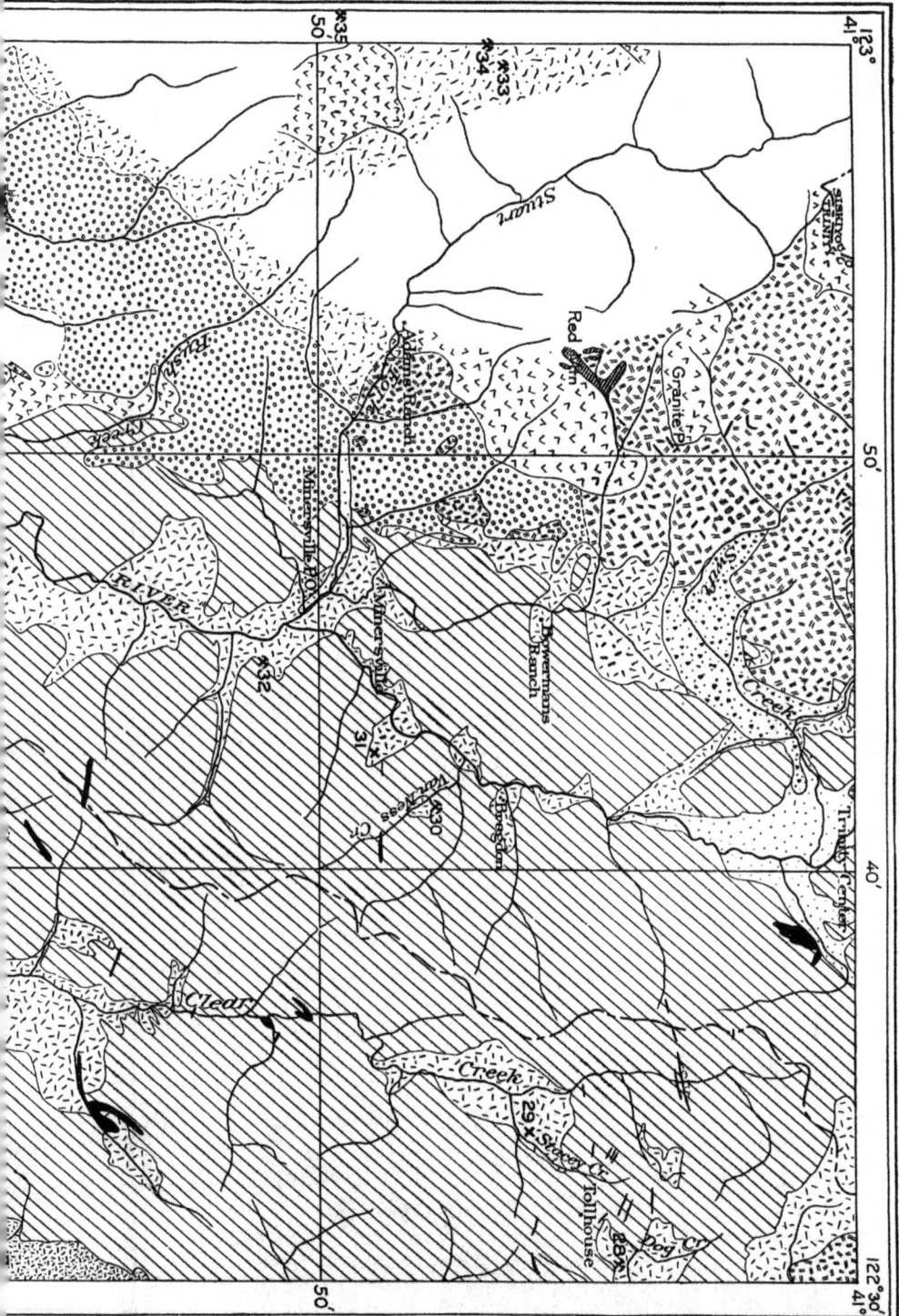

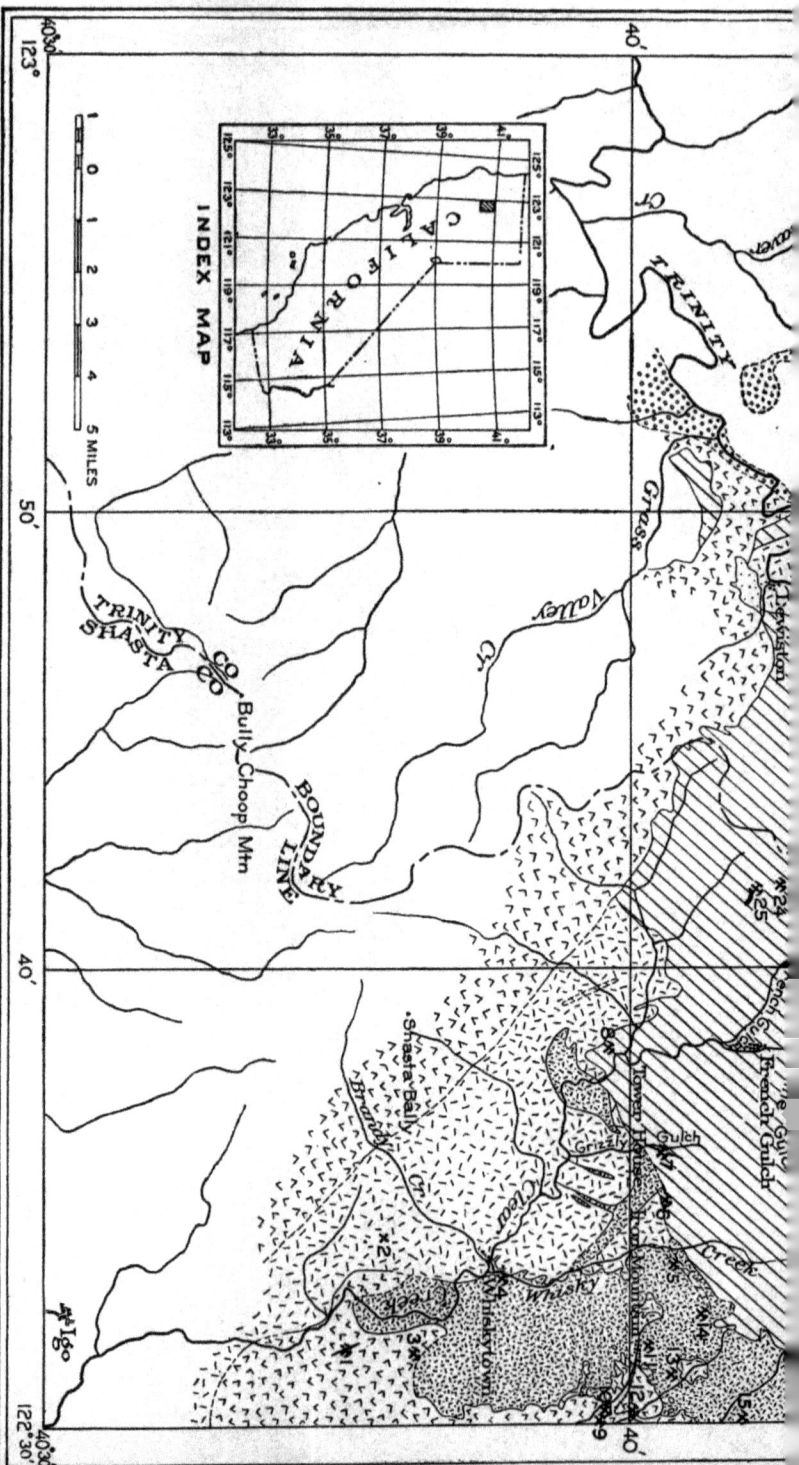

SEDIMENTARY ROCKS

QUATERNARY

 Alluvium

Glacial moraines

TERTIARY

Gravels

CARBONIFEROUS

 Bragdon formation

PRE-DEVONIAN

 Biotite schist with limestone

IGNEOUS ROCKS

LATE JURASSIC OR EARLY CRETACEOUS

 Porphyritic dikes

 Quartz diorite and granodiorite

Alaskite porphyry

Serpentine

DEVONIAN OR OLDER

 Copley meta-andesite

Hornblende schist

 Mine and prospect

INDEX MAP

CALIFORNIA

andesite described below under "Igneous rocks." In the neigh- boring Redding quadrangle[1] the Devonian shale and limestone (Ken- nett formation) are important fossiliferous strata, but they do not ex- tend into this area. Pebbles of fossiliferous Devonian limestone are, however, found in the conglomerates of the overlying Bragdon formation.

The Bragdon formation[2] consists of conglomerates, sandstones, and carbonaceous slates, the slates greatly predominating. The Bragdon rests upon the meta-andesite, but there is evidence of motion almost everywhere along the contact, and only in a few places is anything in the nature of a basal conglomerate to be seen. Here and there a feldspathic sandstone forms the base of the Brag- don. Conglomerates are found in lenticular layers distributed irregularly throughout the formation. These are in places coarse and badly assorted and comprise subangular as well as rounded pebbles. The pebbles consist chiefly of slate, chert, and quartz, with locally numerous limestone pebbles, some of which are fossil- iferous, and rarely pebbles of meta-andesite. The sandstone of the Bragdon is for the most part fine grained. It is dark from the presence of carbonaceous matter and is everywhere more or less feldspathic. Faint cross-bedding is occasionally seen. The shales constitute the major part of the formation and are fine grained and in places extremely carbonaceous. They are locally interbedded with sandstone in thin layers, but more commonly there is a con- siderable thickness of shale beds, uninterrupted by sandstone or conglomerate. Beds of tuff are of rare occurrence. The total thickness of the Bragdon, as estimated by Diller[3] in the Redding quadrangle, is between 2,900 and 6,000 feet. The shale contains numerous fragmentary plant remains. Fossils (chiefly crinoids) which indicate a Carboniferous age have been found by Diller in the conglomerates, sandstones, and tuffs.

Along the contact of the Bragdon with the granodiorite, particu- larly in the vicinity of Lewiston, contact metamorphism has devel- oped a series of highly altered rocks, including quartzite and various types of amphibole and mica schists.

None of the Carboniferous formations which in the Redding quad- rangle overlie the Bragdon formation extend as far west as this area, nor do the Triassic, Jurassic, and Cretaceous sediments of the Redding region enter the surveyed portion of the Weaverville quad- rangle. Tertiary gravels cover a triangular strip west of Trinity River and extend from Trinity Center to the Weaverville basin, and more recent bench gravels follow the present stream courses.

---

[1] Diller, J. S., Redding folio (No. 138), Geol. Atlas U. S., U. S. Geol. Survey, 1906, p. 2.
[2] Hershey, O. H., Am. Geologist, vol. 27, 1901, p. 238; vol. 33, 1904, pp. 248–256. Diller, J. S., Am. Jour. Sci., 4th ser., vol. 19, 1905, pp. 379–387.
[3] Diller J. S., Redding folio (No. 138), Geol. Atlas U. S., U. S. Geol. Survey, 1906, p. 3

The gravels have been studied in detail by Diller.[1]   Local glaciation has carved out cirques and glacial valleys in the Salmon Mountains, and morainal material is found as far into the lowlands as the lower part of Swift Creek.

IGNEOUS ROCKS.

COPLEY META-ANDESITE.

One of the oldest rocks of the area is the meta-andesite, of Devonian age or older,[2] which covers a triangular area of about 6 by 8 miles west of Whiskeytown, and which has been exposed by the erosion of the Bragdon formation in several smaller patches along Clear Creek and Trinity River.

For the most part the formation consists of a series of vesicular lava flows, with minor amounts of tuff and breccia, everywhere much altered. Commonly the rock has a rusty outcrop and a green color on fresh fracture. Intense shearing is the rule, giving a greasy appearance, and it is rare that individual minerals can be distinguished. Where the rock is spherulitic its appearance is extremely characteristic, as small spheres of quartz and epidote 2 or 3 millimeters in diameter are closely crowded together. The breccias are most noticeable on their weathered surfaces, owing to difference in weathering between the rock fragments, which are bleached, and the matrix, which remains a dull green.

In the typical. meta-andesite pyroxene was originally the most prominent mineral, but it is now almost completely altered, usually to chlorite and epidote with a little calcite, more rarely to hornblende. Plagioclase feldspar is present in varying amounts, both in phenocrysts and in the groundmass but is likewise much altered. In many places the rock is composed largely of secondary minerals, chiefly chlorite, epidote, and calcite.

Regional metamorphism has affected the meta-andesite to an extent which makes it impossible to distinguish the individual flows. The spherulitic and breccia phases are most abundant in the region north of Whiskeytown. A specimen from an outcrop of extremely spherulitic lava shows microscopic grains of quartz, apparently original, and hence should be classed as a dacite or rhyolite. Fragments of quartz-bearing lava were also found in one of the breccias. With these exceptions these early lavas seem to be pyroxene andesite.

---

[1] Diller, J. S., The auriferous gravels of the Trinity River basin, California: Bull. U. S. Geol. Survey No. 470, 1911, pp. 11–29.

[2] Diller, J. S., Redding folio (No. 138), Geol. Atlas U. S., U. S. Geol. Survey, 1906, p. 6.

HORNBLENDE SCHIST.

The rock here described as hornblende schist, a part of the Salmon schist of Hershey,[1] outcrops in the western part of the area from near Weaverville north to beyond the Globe mine.   The rock varies greatly in texture and composition.   In the Rush Creek region it is extremely fine grained and the numerous minute hornblende needles give it a silky sheen.   In places the texture is so fine that the individual crystals can not be distinguished with a lens.   Here and there are minute biotite plates (under 2 millimeters in diameter) Under the microscope hornblende is seen to compose from 70 to 90 per cent of the rock.   In places it shows partial alteration to serpentine and chlorite, and the biotite, where present, is chloritized. Small grains of epidote are locally present in association with the hornblende.   A little quartz is always present, either as small lenses parallel to the schistosity or in poorly defined bands with the hornblende.   With the quartz is associated in places a little lime-soda feldspar in small irregular grains.   Magnetite and ilmenite are common accessories.

The coarse-grained type of rock is particularly prominent in the vicinity of the Globe mine.   Here the hornblende is in crystals large enough to be readily identified by the eye and the rock is diversified by small bands and lenses of quartz and feldspar.

The hornblende schist is distinctly older than the other intrusive rocks of the quadrangle, but no data could be obtained as to its age relative to the meta-andesite.

INTRUSIVE ROCKS.

TYPES AND COMPOSITION.

Approximately half the surface of the quadrangle is occupied by intrusive rocks of various types, including serpentinized peridotite and saxonite, quartz diorite, granodiorite, and several types of porphyries and lamprophyres.   The accompanying table illustrates the difference in mineral content of the different varieties.   Their relations, for the most part, have not yet been determined with sufficient certainty to allow any definite statement as to their relative age.   All are intrusive into the Bragdon formation and belong presumably to the same general period of igneous activity.   Diller[2] places the age of the quartz diorite batholith in the Redding quadrangle as late Jurassic or early Cretaceous.

---

[1] Hershey, O. H., Am. Geologist, vol. 27, 1901, pp. 225–245.
[2] Redding folio (No. 138), Geol. Atlas U. S., U. S. Geol. Survey, 1906, p. 8.

Mineral composition of the intrusive rocks of the Weaverville quadrangle.

[0, Lacking; 1, rare or accessory; 2, present in small amount; 3, common; 4, prominent; 5, comprises greater part of the rock.]

Name.	Quartz.	Orthoclase.	Albite.	Soda-lime feldspar.	Biotite.	Hornblende.	Augite.	Monoclinic pyroxene.	Olivine.	Distribution.
Peridotite...............	0	0	0	0	0	0	0	0-2	5	In northeast and northwest corners. Parts of larger areas.
Saxonite................	0	0	0	0	0	0	0	3-4	4	In northwestern portion.
Quartz diorite...........	2-3	0-1	0-1	4	0-2	3	0	0	0	Chiefly in southeastern portion. Part of large batholith.
Granodiorite............	3	2	0-1	3-4	3	0-2	0	0	0	Chiefly in southern and western parts. Blends into granodiorite.
Alaskite porphyry......	3-4	0-1	4	0	0-1	0	0	0	0	Large area in southeastern portion.
Soda granite porphyry..	3	0-1	4	0-1	3-4	0	0	0	0	Numerous dikes and small masses. Cuts the alaskite porphyry.
Diorite porphyry........	0-1	0	0-1	4	0-2	0-3	0-2	0	0	Cuts Bragdon formation in small dikes.
Dacite porphyry........	4	0	0	4	3	2	0	0	0	Found only in two dikes, cutting the Bragdon formation.
Quartz-augite diorite....	1-2	0	0	3	1	0	4	0	0	Dikes, chiefly cutting the quartz diorite.
Lamprophyre (spessartite type).	1	0-1	0	2	2	4-5	1	0	0	Small dikes, comparatively rare.
Lamprophyre (hornblende picrite).	0	0	0	0	0	5	1	0	0	Single dike only.

SERPENTINE.

West of Trinity River a belt of serpentinized basic rock extends for about 8 miles along the northern border of the quadrangle, and reaches southward for about 4 miles. A second and much smaller area lies at the extreme northeast corner of the quadrangle. Both are extensions of larger areas to the north. Chromite has been mined in the serpentine area to the northeast, but none has been found within the quadrangle and no gold deposits occur within the region covered by the serpentine.

The rock of the northeastern area is a peridotite containing everywhere a minor amount of pyroxene. It has a rusty-brown color and a minute lattice-like surface texture on its outcrop but is dark gray, nearly black, on fresh fractures. The cleavage faces of the few pyroxene crystals can be readily distinguished. Pyroxene, largely serpentinized, may form as much as 20 per cent of the volume of the rock. The remainder is composed of olivine, in part altered to serpentine and secondary magnetite.

In the larger area in the northwest corner of the quadrangle much of the rock contains a variable but larger amount of pyroxene (enstatite), so that it is generally a saxonite rather than a peridotite. The alteration product of the pyroxene is more resistant to weathering than the meshlike serpentine formed from the olivine, and hence the weathered surface is very rough and irregular, the pyroxene grains standing out in relief. All the sections examined microscopically are more or less serpentinized, and even in the pyroxene-rich rocks the olivine is somewhat in excess of the pyroxene. Other specimens consist entirely of olivine and its alteration products. The peridotite and saxonite, however, are so intimately mixed that it was not possible to separate them in the field.

QUARTZ DIORITE AND GRANODIORITE.

The southern and western portions of the quadrangle are to a large extent occupied by great masses of granitic rock, parts of a great batholith which underlies a large portion of the Klamath Mountains. A small area south of Whiskeytown is an extension from the larger mass mapped by Diller[1] in the Redding quadrangle. To the west of this area, separated only by a narrow strip of meta-andesite cover, is the rock which forms the mass of Shasta Bally, and farther northwest are isolated patches in the Salmon Mountains. In the lowlands the rock is deeply weathered, owing to the decomposition of the dark silicates, and in places is reduced to a quartz and feldspar sand. On the steeper slopes, as on Shasta Bally and on the flanks of Red Mountain, where erosion is rapid, the less weathered portions of the rock stand out as great monoliths. In the more rugged canyons and glaciated portions of the Salmon Mountains fresh granodiorite forms prominent cliffs, whose faces follow the steeply inclined joint planes.

The rock exhibits a considerable variety in both textural and mineralogic characteristics. Its most usual type is a medium-grained (2 to 4 millimeters) granular rock carrying quartz, plagioclase feldspar, little or no orthoclase, and both biotite and hornblende. As a rule it is even grained, but in places, as along the toll road east of Whiskeytown, it is distinctly porphyritic, with large corroded phenocrysts of quartz. Aplitic dikes are not uncommon, and many hornblendic streaks (schlieren) and irregular segregations rich in hornblende are found near the borders of the mass.

The ferromagnesian minerals, hornblende and biotite, vary greatly in their relative proportions; in certain parts of the area biotite is dominant and hornblende is present in minor amount or completely lacking, while elsewhere hornblende is the only essential dark silicate present. The greater part of the feldspar individuals are plagioclase and many of them show a zonal structure. The usual composition is oligoclase or oligoclase-andesine with a center of andesine. A large proportion of the feldspars are untwinned and in some of the slides examined these are in part orthoclase. Although no thorough study of the feldspars has been made it is believed that even where orthoclase is most abundant it rarely exceeds 10 per cent of the rock. Orthoclase appears to be most abundant where biotite is the dominant ferromagnesian mineral. Quartz is present in small anhedral grains except in the porphyritic phase already referred to, where it occurs

---

also as corroded phenocrysts a centimeter or more in diameter. Accessory minerals are inconspicuous. They include apatite, magnetite, ilmenite, zircon, and, in one specimen, garnet.

ALASKITE PORPHYRY.

Alaskite porphyry [1] is a rock of great economic importance in the extreme eastern part of the quadrangle and in the region to the east, as it forms the wall rock of the Shasta County copper deposits. In the Redding quadrangle it covers a large area. In the Weaverville quadrangle it lies along the eastern border in an irregular north-south band some 13 miles in length and about 3 miles wide. In its northern portion it was intruded between the Bragdon formation and the Copley meta-andesite and in the south it separates the quartz diorite and the meta-andesite. A small discrete area lies to the south of Tower House, and dikes were encountered on Clear Creek north of Cline Gulch and on Van Ness Creek southeast of the Five Pines mine.

The rock varies greatly in texture. A coarse-grained porphyritic type that outcrops along the eastern boundary of the area closely resembles the porphyritic quartz diorite, into which it grades. The finer-grained type is more usual and is particularly well exposed in the region north and northeast of Whiskeytown. It is white or light green in color, gray in the freshest specimens, and where pyritized is heavily iron stained on the outcrop. In outward appearance it resembles a rhyolite, as it has a distinctly platy fracture, due to later shearing, and shows minute crystals of quartz and more rarely of feldspar in a dense aphanitic groundmass.

The distinguishing feature of the mineral composition of the alaskite porphyry is the absence of ferromagnesian minerals. Even in the freshest specimens, microscopic examination showed no dark silicates except small specks of chlorite, which possibly represent altered biotite, and minute grains of epidote and zoisite, which may likewise indicate the original presence of a small amount of ferromagnesian minerals. The feldspar, both the phenocrysts and microlites in the groundmass, is albite. The groundmass, even of the most aphanitic type, is holocrystalline and consists of an aggregate of minute grains of quartz and rods of feldspar.

SODA GRANITE PORPHYRY.

Soda granite porphyry is closely allied to the alaskite porphyry in mineral composition. Dikes of this rock have been noted by Butler [2] cutting the alaskite porphyry in the Iron Mountain district. Similar

[1] Graton, L. C., The occurrence of copper in Shasta County, Cal.: Bull. U. S. Geol. Survey No. 430, 1910, p. 81.

[2] Butler, B. S., Pyrogenetic epidote: Am. Jour. Sci., 4th ser., vol. 28, 1909, p. 27.

dikes are common throughout the Bragdon formation and appear to have a rather close association with the fissure veins in that formation. The rock is always porphyritic but varies greatly in texture in different localities. Most commonly the phenocrysts are about equal in volume to the groundmass, and the groundmass, though in places extremely fine grained, is everywhere distinctly crystalline. The phenocrysts named in the usual order of their abundance are biotite, feldspar, and quartz. Biotite occurs generally in thin hexagonal plates commonly not over 2 or 3 millimeters in diameter. The feldspar phenocrysts are euhedral but generally very small, few of them exceeding 3 millimeters in length. Most of the determinable feldspars proved to be nearly pure albite. In a few dikes, however, the prevailing feldspar is oligoclase, and more rarely a little orthoclase is present. Quartz crystals are generally less prominent than biotite and feldspar, but the relative amount varies greatly in the different dikes. Few of the phenocrysts show good crystal outlines, and deep embayments due to magmatic corrosion are a characteristic feature. The accessory minerals are titanite, zircon, magnetite, and rarely colorless garnet and epidote.[1] The groundmass is always holocrystalline but in most places extremely fine grained; it consists of quartz and feldspar with scattered grains of the accessory minerals. The feldspar of the groundmass, in most of the slides examined, seems to have the same composition as the feldspar phenocrysts.

DIORITE PORPHYRY.

Diorite porphyry, locally known as "bird's-eye porphyry," forms prominent dikes in the Bragdon formation, particularly in the French Gulch and Whiskeytown region. The noticeable feature of the rock is the presence of numerous white feldspar phenocrysts, some of them as much as a centimeter in length, in a dark-gray groundmass. The composition of all the feldspars was not determinable, but in general they vary between albite-oligoclase and calcic andesine, the latter being the most common. Hornblende is the most abundant ferromagnesian mineral and may generally be seen in the hand specimen as small needles 1 or 2 millimeters long. Augite or biotite may be present as well as hornblende, but neither is common and augite is the rarer of the two. In one dike, however, augite is the only dark silicate present and exceeds the feldspar phenocrysts in volume. Small biotite plates are more common, and in a dike near the Washington mine biotite is the only ferromagnesian mineral present. The groundmass is extremely fine grained and consists of very minute feldspar laths with rare specks of hornblende and biotite.

---

[1] Butler, B. S., Pyrogenetic epidote: Am. Jour. Sci., 4th ser., vol. 28, 1909, p. 27.

DACITE PORPHYRY.

Dikes of dacite porphyry cut the Bragdon formation at Smiths Gulch, 4 miles north of French Gulch, and on the East Fork of Clear Creek. Specimens from the dike at Smiths Gulch have been included in the educational series of rock specimens of the Geological Survey and have been described in detail by J. P. Iddings.[1] The prominent feature of the rock is the presence of large rounded quartz phenocrysts, the largest nearly a centimeter in diameter. Phenocrysts of milky-white oligoclase feldspar are more numerous than those of quartz. Small biotite and hornblende crystals are also common.

Rock of this type forms several dikes in the Redding quadrangle and appears to correspond to the granodiorite porphyry of the Headlight mine, north of Trinity Center, described by D. F. MacDonald.[2]

QUARTZ-AUGITE DIORITE.

Dikes of quartz-augite diorite occur in the granite area south of Whiskeytown, near the Mount Shasta, Mascot, and Gambrinus mines, and on Clear Creek 9 miles north of French Gulch. There is a large intrusion of a rock of similar composition in the central part of the Redding quadrangle.[3] The rock also appears to be essentially like the basaltic dikes which MacDonald[4] considers to be genetically connected with many of the ore deposits of the region immediately north of the Weaverville quadrangle.

The rock is dark colored, as the dark silicates exceed in amount the quartz and feldspar. Wherever seen it is of granular texture, the grains being about 1 millimeter in size. Augite is the dominant mineral but is largely chloritized. It is possible that a little biotite was also originally present. The feldspar is much altered, but wherever determinable proved to be andesine. Original quartz occurs in small amount and is interstitial between the augite and feldspar.

LAMPROPHYRIC DIKES.

A wide variety of basic dikes has been found in the quadrangle, particularly along the borders of the granodiorite masses. They are small and inconspicuous and as a rule do not exceed a few feet in width. Hornblende is the most prominent mineral in both the phenocrysts and the groundmass and may form as much as 70 per cent of the volume of the rock. Biotite is the second ferromagnesian mineral in abundance, though always far less in amount than hornblende. In one dike, however, biotite is lacking and in its place there is a small amount of augite. Plagioclase feldspar, from calcic

---

[1] Bull. U. S. Geol. Survey No. 150, 1898, pp. 233–236.

[2] Bull. U. S. Geol. Survey No. 530, 1913, p. 13.

[3] Diller, J. S., Redding folio (No. 138), Geol. Atlas U. S., U. S. Geol. Survey, 1906, p. 8.

[4] MacDonald, D. F., Gold lodes of the Carrville district, Trinity County, Cal.: Bull. U. S. Geol. Survey No. 530, 1913, p. 14.

andesine to labradorite, forms a large part of the groundmass. Quartz occurs as a minor constituent in two of the dikes, and a small amount of orthoclase appears to be present in a third. Chlorite, epidote, and calcite are prominent as alteration products. The mineral composition of the rock appears to agree closely with that of the dike rock spessartite, as defined by Rosenbusch.[1]

A very rare type of dike rock is hornblende picrite, which consists essentially of hornblende.

GOLD DEPOSITS.

PRODUCTION.

Gold was first discovered in the Weaverville quadrangle in 1848; Raymond[2] states that in the fall of that year Maj. Redding took out $60,000 worth from the bed of Clear Creek. Lode mining began in 1852 with the location of the Washington mine in the French Gulch district,[3] but for many years the output from the lodes was far below that of the rich gulches and bench gravels. Lode discoveries were constantly being made, however, and the waning importance of placers in recent years has been in part compensated by increased lode production. It is impossible to make any close estimate of the amount of gold produced from the lode mines of the quadrangle, but from the fragmentary data available it is believed that the total is in excess of $15,000,000.

The following table shows the annual gold and silver production of Shasta and Trinity counties. It is believed that from 30 to 60 per cent of the lode production of each county is derived from mines within the Weaverville quadrangle.

Value of gold and silver from lode mines in Shasta and Trinity counties, Cal.

[From reports of the Director of the Mint.]

	Shasta County.	Trinity County.		Shasta County.	Trinity County.
1897	$634,632	$355,773	1901	$1,799,578	$360,237
1898	1,014,633	281,055	1902	906,283	330,785
1899	1,050,023	219,653	1903	957,602	275,342
1900	1,357,350	263,939			

[1] Rosenbusch, H., Mikroskopische Physiographie der massigen Gesteine, Stuttgart, 1907, p. 681.

[2] Raymond, R. W., Mining in the States and Territories west of the Rocky Mountains, Washington, 1874, p. 143.

[3] Trask, J. B., Report on the geology of northern and southern California: Rept. California State Geologist, Sacramento, 1856, p. 49.

22652°—Bull. 540—14——3

Production of gold lode mines in Shasta and Trinity counties, Cal., in fine ounces, except as indicated.

[From Mineral Resources of the United States.]

Year.	Shasta County.			Trinity County.		
	Gold.	Silver.	Number of producing mines.	Gold.	Silver.	Number of producing mines.
1903	a 34,462	a ($214,028)	....	14,038	($184)	....
1904	31,024	($5,096)	b 35	9,165	($80)	27
1905	27,512	8,025	25	17,020	3,759	30
1906	27,133	8,817	17	7,119	1,881	16
1907	23,986	42,465	23	8,713	1,271	19
1908	34,056	66,362	28	8,980	4,057	24
1909	47,532	27,279	28	7,342	2,144	34
1910	44,293	8,752	34	6,134	1,592	24
1911	38,193	23,585	49	14,411	10,262	24

a Includes gold and silver production from Shasta County copper mines.
b Includes copper mines.

TYPES.

The principal gold deposits of the quadrangle are fissure veins, as a rule narrow, with steep dips. Certain minor deposits known as "pockets" form a distinct type.

The fissure veins are most numerous in the slate, are more rare in meta-andesite near the slate, and in both formations are usually associated with porphyry dikes. Deposits of this type have been the best producers of the quadrangle. A second type of fissure veins comprises those which cut the quartz diorite and alaskite porphyry, particularly in the region south and east of Whiskeytown. As a rule, basic dikes occur in the vicinity of these deposits. In the Dedrick district the fissure veins have walls of hornblende schist, and granitic and alaskitic dikes occur in the vicinity. The pocket deposits are always on or near the contact of the slate of the Bragdon formation with another rock, generally meta-andesite.

DISTRIBUTION.

FISSURE VEINS IN THE BRAGDON FORMATION.

Most of the fissure veins in the Bragdon formation lie in the vicinity of the complex of porphyritic intrusives which extends from the upper part of French Gulch across the divide into Trinity County. The following table shows the principal features of the more important mines of this group, and brings out the similarity in the occurrence of the veins and their mineral content:

Principal characteristics of the most important fissure veins in the Bragdon formation.

Mine.	No. on map.	Production	Wall rocks.	Vein.	Mineralogy.	Remarks.
Whiskeytown district.						
Truscott	7	Estimated $60,000, 1887–1912.	Slate and diorite porphyry.	Strike N. 20° E.; dip 60°–80° S. Lenticular, maximum width 10 feet. Follows contact.	Quartz, calcite, pyrite, and rare chalcopyrite. Tellurides reported.	Also a small rich stringer 8 inches wide in porphyry.
French Gulch district.						
Gladstone	18	$2,500,000, 1896–1912	Slate and sandstone; soda granite porphyry in workings.	Strike east; dip vertical and steep north; local irregularities. Average width, 2½ feet.	Quartz with small amount of calcite, pyrite, galena, sphalerite, and arsenopyrite rare.	Vein developed to a depth of 2,000 feet below outcrop.
American	19	Unknown	Slate and sandstone	Strike N. 80° E. to east; dip 80° S. to vertical.	Quartz, pyrite, and arsenopyrite.	Not working. No intrusive rocks seen in immediate vicinity.
Franklin	20	$350,000, 1907–1912	Slate and soda granite porphyry.	Two veins: 1. Strike N. 5° to 30° W.; dip steep east. 2. Strike east; dip 70° N.	Quartz, calcite, pyrite, arsenopyrite, galena, and rare sphalerite.	Veins are near contact but cut both rocks.
Washington	22	Estimated between $1,000,000 and $2,000,000, 1852–1912.	Slate and meta-andesite. Many intrusives, soda granite porphyry, quartz-augite diorite, and diorit porphyry in the vicinity.	Two veins: 1. Strike north; dip 65° E. 2. Strike east; dip steep north.	Quartz, no calcite, pyrite, galena, rare arsenopyrite, and sphalerite.	Veins cut slate and meta-andesite. Not controlled by contact.
Niagara	23	Unknown; probably about $1,000,000.	Slate, conglomerate, diorite porphyry, and soda granite porphyry on dump.		Quartz, pyrite, galena, rare sphalerite, and arsenopyrite.	Not in operation.
Summit	24	Estimated $200,000	Slate and soda granite porphyry.	Two veins, crossing dike.	Quartz, very small amount of calcite, pyrite, galena, sphalerite, and arsenopyrite, manganese oxide.	Ore partly oxidized. Vein productive only in porphyry. Ore pockety and very rich in spots.
Brunswick	25	$70,000, 1879–1912	Slate and diorite porphyry.	Strike east; dip 60° N.	Quartz with small amount of calcite, pyrite.	Vein follows contact rather closely.
Deadwood district.						
Brown Bear	26	Estimated $7,000,000 to $10,000,000, 1875–1912.	Slate, diorite porphyry and soda granite porphyry.	Two principal veins. Strike N. 80° E.; dip north and south.	Quartz, minor calcite, pyrite, galena, sphalerite, and rare arsenopyrite.	Veins generally close to contacts. Complex of porphyry dikes and irregular masses.

Principal characteristics of the most important fissure veins in the Bragdon formation—Continued.

Mine.	No. on map.	Production.	Wall rocks.	Vein.	Mineralogy.	Remarks.
Dog Creek district.						
Delta and Trinity.	28	$32,000	Meta-andesite, alaskite porphyry, and soda granite porphyry; slate short distance above.	Several small veins. Strike between east and N. 70° E.	Quartz, minor calcite, rare barite, pyrite, galena, sphalerite, and rare arsenopyrite and chalcopyrite. Manganese oxide.	
Minersville district.						
Fairview.	32	Estimated $200,000.	Slate; meta-andesite near by.	Strike west to N. 80° W.; dip 50°-80° N. Maximum width 20 feet.	Quartz, pyrite.	Ore from $7 to $30 a ton. No intrusive seen in vicinity.

The slate wall rock is generally much crushed and sheared, so that the bedding has been obliterated and the rock altered to a glistening black slate with a shining luster and markedly schistose structure. This slickensiding has emphasized the presence of carbon in the slate by concentrating it along the shear planes, locally in sufficient amount to blacken the hands. The dikes, however, show no such shearing. Intrusive porphyries, either soda granite or diorite porphyry, are found in close association with almost all the veins. Where the intrusive is diorite porphyry, in the Truscott and Brunswick mines, the veins follow the contact of slate and porphyry. In the Summit mine the vein cuts directly across the soda granite porphyry dike. In the Brown Bear and Franklin mines the veins, though near the contact, cut both the soda granite porphyry and the slates, but tend to be best developed in the porphyry and to finger out in the slates. In the Gladstone and Washington mines, on the other hand, the veins are in the slate and do not cut the soda granite porphyry. This leads to the conclusion that the position of the veins along the contact of the diorite porphyry is merely dependent on the fact that the contact of two such unlike rocks is a favorable position for fissuring, and that the ore deposition is genetically connected with the intrusion of the soda granite porphyry.

The mineral composition is much the same in all the veins. The gangue is quartz with a minor amount of calcite and small specks of green mica. Certain sulphides, notably pyrite, galena, sphalerite, and arsenopyrite, are fairly constant, while chalcopyrite is very rare. To a certain extent the sulphides vary with the wall rock. Where the walls are soda granite porphyry, arsenopyrite is always found both in the ore and in the country rock. Galena and sphalerite are more common where the walls are slate. Much of the gold is coarse enough to be easily seen by the naked eye. Manganese oxide is not a prominent feature of any of the veins of this type, though in the oxidized ores, such as are mined in the Summit and the upper workings of the Washington, the rich pockets are marked by the sooty black powder. The surface ores in many deposits were known to have been exceedingly rich, but have now been largely exhausted.

The veins are fairly persistent both in strike and in dip. The Last Chance and Monte Cristo veins of the Brown Bear mine have been drifted for a distance of 1,400 feet and the Gladstone vein for 2,000 feet. The Gladstone vein, moreover, has been developed for 2,000 feet vertically below the outcrop. The ore shoots are generally large and fairly regular, the maximum drift length being about 500 feet. The lowest point reached in the Gladstone shaft is about 800 feet above sea level, and the veins of the Niagara group outcrop on the Trinity-Shasta divide, 7 miles to the west, at an elevation of about 4,200 feet, which gives a minimum vertical range of deposition of

3,400 feet. The veins (in meta-andesite below the slate) of the Dog Creek district have been included in this group because of their mineralogic similarity but are less persistent in strike and have not yet been explored to any great depth.

The prevailing trend of the fissures is east and west, though with many minor variations. In the western part of the French Gulch district there is also a minor north-south series.

FISSURE VEINS OUTSIDE OF THE SLATES.

Most of the important veins outside the slate area are in the Iron Mountain, Igo, and Bully Choop districts, which were not visited in 1912. In the Whiskeytown and Dedrick districts the veins are poorer in sulphides and show a smaller variety of minerals than those which cut the Bragdon formation. In several veins pyrite is the only sulphide present. Quartz is the principal gangue mineral and calcite is comparatively rare.

In the mines of the Whiskeytown district small dikes of an augite-quartz diorite, with a very large proportion of dark minerals, are near the veins. In two of the mines, the Mascot and Gambrinus, the altered wall rock is said to carry gold.

Principal characteristics of fissure veins outside of the Bragdon formation.

Mine.	No. on map.	Production.	Wall rocks.	Vein.	Minerals.	Remarks.
Whiskeytown district.						
Mount Shasta	1	$178,000, 1897-1911	Alaskite porphyry	Two veins; strike north-west; dip varying.	Quartz, rare calcite, pyrite, molybdenite; no free gold.	Ore from $4 to $43 a ton. Dike of basic quartz-augite diorite near vein.
Mascot	3	None	Basic quartz-augite diorite, near quartz diorite contact.	Two veins; 1. Strike N. 55° E.; dip 45° SE.; 2. Strike N. 50°-62° E.; dip 60° SE.; width 6 inches to 3 feet.	Quartz, pyrite, manganese oxide; ore largely oxidized.	Ore said to run $11.85 a ton; quartz diorite grades into alaskite porphyry a short distance to the north.
Gambrinus	4	$127,000, 1870-1912	Alaskite porphyry, near meta-andesite contact.	Four veins; strike N. 50° W. to west; dip 50° N. to vertical.	Quartz, pyrite, rare chalcopyrite, manganese oxide, rare albite.	Altered porphyry said to be auriferous. Basic quartz-augite diorite dike near by.
Mad Ox	5	Unknown	Meta-andesite, near alaskite porphyry intrusion.	Strike N. 22°-33° E.; dip 80° SE. to vertical; width from knife-edge to 4 feet.	Iron-stained quartz, rare calcite; no sulphides seen.	Mine not working. Ore oxidized.
Dedrick district.						
Globe	33		Hornblende schist	Strike N. 55°-70° E.; dip 60° SE.; average width 8 feet.	Quartz, albite, rare calcite, pyrite, manganese oxide.	Soda granite and alaskite porphyry dikes near by. Ore about $10 a ton.
Craig	35		...do...	Strike N. 70° E.; dip 45°-60° SE.	Quartz, rare calcite, pyrite, rare chalcopyrite.	Gold more finely divided than usual.

POCKET DEPOSITS.

In certain of the deposits of the region practically all the gold is contained in small scattered pockets near the surface, all of them being found at or near the contact of the black slate with some other formation, generally meta-andesite. At three localities, the Eldorado, Mad Mule, and Five Pines mines, these pockets have been found of sufficient size or close enough together to justify extensive work along the contact. Elsewhere they have been explored for only a few feet below the surface. Of the three mines, two, the Eldorado and the Five Pines, follow the contact of the slate and meta-andesite, whereas the pockets of the Mad Mule mine (fig. 3, p. 47) lie in troughs formed by irregularities in a dike of diorite porphyry.

The Eldorado perhaps represents an intermediate type between the fissure veins in the Bragdon formation and the typical pocket deposits. In this mine manganiferous quartz lies along a faulted contact between meta-andesite and slate, with more or less gouge on both walls. The movement, although following the contact in a general way, has been in part oblique to it. Hence the present line of contact is extremely serrate; sharp wedges of glistening, slickensided black slate enter the meta-andesite, and small lenticular masses of slate are completely cut off from the main body. The pockets, from one of which $2,500 worth of gold was taken, are found either in the gouge on the walls of the manganiferous quartz or in little irregular veinlets of quartz in the meta-andesite close to the inliers of slate. So far work has been profitable only in the upper levels. The fineness of the gold, however, is notably low, as $14 an ounce is said to be the average value.

In the Five Pines mine the pockets likewise follow a faulted slate and meta-andesite contact, but here they lie downhill from a vein of low-grade manganiferous quartz which crosses the contact and near irregular stringers of quartz in the meta-andesite near the slate. Along the contact of slate and meta-andesite are patches of calcite with a small amount of quartz, irregularly mixed with black slate. The pockets, of which one yielded as much as $45,000 in a distance of 44 feet, are found along the slate contact. Less commonly small pockets are found in the meta-andesite near the slate, and more rarely still in small calcite stringers in the slate close to the contact. Near several of the larger pockets small discontinuous quartz stringers cut the porphyry. The gold, however, is commonly between the calcite and slate or along cleavage planes in calcite. Pockets have been found from the surface to the water level, and always along watercourses.

In the Mad Mule mine the pockets lie along the contact of the slate with a steeply dipping dike of diorite porphyry, here again in connection with patches of calcite. The porphyry is somewhat pyritized and is cut by small quartz stringers that carry much manganese oxide.

These are looked upon as indicators of rich pockets. The slate is much sheared, especially along the contact, and as in the Five Pines mine, the pockets lie along the present watercourses.

A deposit of similar type in the vicinity of Minersville has been noted by Diller.[1] Here gold occurs in calcite lenses in black slate, which also carry a small amount of quartz. In one hand specimen of the ore three-fourths of its volume is estimated to be native gold.

The most common type of pocket deposit, however, is that in which the gold has been followed only a few feet below the surface. These pockets are found exclusively along the contacts of slate and meta-andesite, particularly in the region between Trinity Center and Minersville. Much of the early placer mining consisted in sluicing these rich surface pockets. This type of deposit has been studied by Hershey,[2] who reaches the conclusion that the pockets are the result of solution of the gold contained in pyritized zones in the meta-andesite by surface waters, and redeposition by contact with the black slate. He says, in part:

At the contact the black rock frequently has a shining luster and a schistose structure due to shearing. This gives it imperfectly the power of a gouge to deflect underground waters. The volcanic rock near the contact has generally been decomposed, softened, and changed to a dull-brown color; it is popularly known as porphyry. In places there is a thin vein of quartz between the so-called porphyry and the black schistose material, but generally they are in actual contact or separated merely by a thin seam of ferruginous dirt. The dirt seam often carries a little free gold, but the pockets are said to be found near or where seams of quartz penetrate the porphyry downward from the contact. The gold lies in a thin, flat sheet upon the igneous rock and under the slate, and in some cases extends a short distance into the former formation, rarely into the latter. It is in the form of coarse and fine grains that have a peculiar smooth and rounded surface quite unlike the free gold in quartz veins. * * *

The reason that the slate-volcanic rock contact is the great "pocket" horizon is that it is there that the gold-bearing solution first reaches a carbonaceous rock—the carbon precipitates the gold. The water may reach the contact by traveling nearly horizontally through inclined strata or by ascending under hydrostatic pressure. The sheared slate so frequently found along the contact aids in holding the solution to it while the gold is being deposited. Probably also water issuing from the slate carries the precipitating agent. For a long time the point of union between the precipitant and the gold-bearing water remains at one place at or near the contact, and thousands of dollars' worth of gold is thrown down within a space of a few cubic yards or less.

It seems clear that these pocket deposits, including both those of the three mines already mentioned and the smaller deposits, are of sur-ficial origin and that the factor determining the deposition of the gold is the carbon of the black slates, as has been stated by Hershey. It is believed, however, that the cause of the solution of the gold is the presence of manganese oxide. Manganiferous quartz is present in the three mines of the pocket-deposit type. The writer was not fortunate enough to see any of the smaller pockets that are being

---

[1] Diller, J. S. Native gold in calcite: Am. Jour. Sci., 3d ser., vol. 39, 1890, p. 160.

[2] Hershey, O. H., Origin of gold pockets in northern California: Min. and Sci. Press, vol. 101, 1910, pp. 741-742.

worked, but in all the old workings examined joints in the meta-andesite near the slate contact were in most places stained with manganese oxide.

W. H. Emmons [1] has discussed the enrichment of gold deposits in veins in igneous rocks. In the deposits considered ferrous sulphate is the precipitant, and, owing to its continual oxidation to ferric sulphate through the presence of manganese oxide, it does not precipitate gold to any extent until the water level is reached. It may be assumed, however, that in these deposits the precipitant has been the carbon of the black slates, whose efficiency is unaffected by its being in the oxidized zone, and perhaps the calcite of the small lenses as well. Moreover, as the calcite, which is present in the larger of these deposits, is dissolved, the acid solution is neutralized and becomes no longer capable of taking gold into solution. Hence deposition of gold by surface waters is here confined to a narrow zone close to the surface, and where favorable local conditions control the flow, as has been most clearly the case in the Mad Mule mine, pocket deposits of small size but extraordinary richness may be formed. In the Mad Mule a single plate of gold weighing 100 ounces is said to have been found between the calcite and the slate.

That gold is not dissolved to any extent through the agency of manganese in the presence of calcite, owing to the neutralization of the acid water by the solution of calcite, is further illustrated by studies made by Eddingfield [2] on certain gold-calcite-manganese ores of the Philippine Islands.

For the formation of such deposits as the Mad Mule the first step seems to have been the fissuring near the slate contact and the formation of small veinlets of gold-bearing manganiferous quartz, some of them carrying calcite. At about the same time came faulting along the contact, with the resultant slickensiding, concentration of the carbon of the slate along the slickensided surfaces, and the formation of gouge, with filling of lenticular open spaces along the contact by primary solutions that deposited chiefly calcite but also some quartz, pyrite, and arsenopyrite. The formation of the present pockets did not begin until a topography approaching that of the present time was attained. The course of the underground water is to a large extent controlled by the slickensided slate along the contacts and by joint planes in the more massive meta-andesite or porphyry. These waters are acid owing to the decomposition of the pyrite in the igneous rock. Whatever calcite may have been present in the manganiferous quartz stringers was soon dissolved and possibly redeposited on the already formed calcite lenses along the contacts. Later acid waters, following the same channels and no longer neutralized by the calcite, were able, through the agency of the manganese

[1] Trans. Am. Inst. Min. Eng., vol. 42, 1911, p. 3.

[2] Eddingfield, F. T., Philippine Jour. Sci., vol. 8, sec. A, 1913, pp. 125-134.

oxide, to dissolve the gold carried in the manganiferous quartz stringers or in the pyritic bands in the meta-andesite,[1] but not to transport it to any great distance, owing to the precipitating action of the carbon of the black slate. Nor can the gold thus deposited be again readily dissolved, owing to the neutralization of the solution by the calcite of the lenses along the contact. Thus a shallow zone of rich deposits is formed, the lower level of which can never be far from the original source of the gold. Rich placer deposits have been the rule wherever the streams cut the slate and meta-andesite contact.

MINERALOGY.

The primary minerals mentioned below have been noted in the deposits.

GANGUE MINERALS.

Quartz is the principal gangue mineral of all the fissure veins. It is rarely drusy, and as seen under the microscope it is generally in small interlocking grains, filled with minute inclusions. Where small fragments of the wall rock have been replaced the grain is much finer.

Calcite is almost universally present in minor amounts and is the principal mineral of the pocket deposits. Its brown color on weathered surfaces is evidence of the presence of manganese. In the Five Pines mine the fresh calcite has a distinct pinkish tinge, presumably due to a high manganese content.

Mica in the form of paragonite or sericite is common as an alteration product in the altered porphyry close to the veins. In the quartz of a few of the veins, generally close to the walls, are small specks of a dark-green mineral which appears to be muscovite, probably the variety mariposite. Under the microscope a few minute veinlets of quartz and mica were seen crossing pyrite crystals in the ore.

Albite occurs as a minor gangue mineral in the quartz of the Globe vein and was observed in microscopic grains in the ore of the Gambrinus mine. Specimens of vein quartz from the Mount Shasta mine contain kaolin, which may have resulted from alteration of original feldspars.

Barite in small tabular crystals was observed in ore from the Five Pines mine and the Red Lion claim of the Delta mine. In the Five Pines mine it is associated with pink calcite; on the Delta property it appears in part to replace alaskite porphyry and is cut by small veinlets of pyrite.

Tourmaline was found only in the ore of the Mountain Monarch prospect, in small rosettes 4 or 5 millimeters in diameter, composed of minute acicular crystals.

---

[1] Hershey, O. H., op. cit., p. 742.

METALLIC MINERALS.

Arsenopyrite is common in nearly all the mines in the Bragdon formation. As a rule, however, it is found in the altered porphyry rather than in the ore itself and in the neighborhood of the porphyry rather than in that of the black slate.

Chalcopyrite is a comparatively rare mineral in the gold deposits but was noted in the quartz of the Gambrinus and Craig gold mines and, together with pyrite, in the Mountain Monarch and Delta copper prospects.

Covellite was seen only as a coating on some of the pyrite in the ore of the Mountain Monarch prospect.

Galena is an important mineral in the deposits in the veins in the slate but is not seen elsewhere. Its presence in the quartz is regarded as a sign of rich ore. It is more common in the quartz close to the black slate than near the porphyry wall rocks and is never found outside the vein.

Gold occurs in the veins both as free gold and in the sulphides. The free gold is by far the most important, as the concentrates do not often exceed 1 per cent of the ore in weight or 6 per cent in value. In many of the deposits the concentrates are not considered worth saving, and the pocket deposits contain practically no sulphides. In many of the fissure veins in the slate the gold is in plates large enough to be easily visible. Most commonly it is closely associated with included specks of black slate or occurs near the slate wall rock, and in a number of places it has been deposited close to galena, or more rarely sphalerite, in the quartz. In rich oxidized ores, such as those of the Washington surface workings or the Summit mine, gold is commonly present in small cavities in the quartz associated with iron or manganese oxide. In the pocket deposits gold may be found in irregular plates between the slate and meta-andesite or associated with manganese oxide in joint planes of the meta-andesite, but in the larger deposits of this type, such as the Mad Mule and the Five Pines, it also occurs along cleavage planes of calcite or between calcite and slate.

Manganese oxide is found in the pocket deposits and in most of the fissure veins, though it is less prominent in the veins in the Bragdon formation than in several of the others.

Molybdenite occurs in small specks in the ore of the Mount Shasta mine.

Pyrite is the most widespread of all the metallic minerals. It is found in the quartz of the fissure veins and calcite lenses, in the shear planes and small fissures in the slates, and in small crystals scattered through the porphyries, even at a distance from the vein.

Sphalerite occurs only in the veins which cut the slate and is nowhere prominent. Generally it is to be found in close association with galena.

Tellurides have been reported from a few fissure veins, but no indication of telluride minerals was seen in any of the specimens examined.

The two important gangue minerals, quartz and calcite, seem to have crystallized simultaneously. In many veins quartz is on the walls and calcite in the center, but this relation is not at all constant. In some specimens of oxidized ore the calcite has been dissolved out, leaving casts of tabular crystals impressed on crystalline quartz, showing an intergrowth of the two minerals, and certain specimens show veining of each of these minerals by small stringers of the other. Of the sulphides arsenopyrite and pyrite have migrated into the wall rock to a certain extent, but galena and sphalerite are found only in the quartz. The gold has been deposited chiefly near the black slate and to a less extent on the galena and sphalerite. Calcite is nearly always free from sulphides but is cut by small stringers of quartz, which in one specimen carry sulphides, and many of the pyrite crystals are cut by microscopic veinlets of quartz and sericite.

HYDROTHERMAL ALTERATION.

The country rocks adjoining the veins show comparatively little alteration that is directly attributable to the ore-bearing solutions. Where the walls are slate small fragments in the vein are replaced by an aggregate of quartz, sericite, pyrite, and, rarely, arsenopyrite grains. The carbonaceous matter remains unchanged and locally forms a nucleus for the deposition of minute flakes of gold. The quartz is distinctly finer grained than in the veins. Sandstone near the veins is impregnated with pyrite, and its feldspar grains are in part sericitized. Where the soda granite porphyry is in immediate contact with the vein the chief alteration consists in silicification, with the development of pyrite and arsenopyrite. At a distance from the vein the alteration consists in the sericitization of the feldspar, with the development of a small amount of secondary calcite and the chloritization and, to a minor degree, sericitization of the biotite. Chlorite and small specks of epidote are often found in the altered wall rocks, but these minerals are not especially prevalent in the vicinity of the veins.

Studies by B. S. Butler [1] have shown that near the copper deposits a part of the micaceous secondary mineral is paragonite rather than sericite. As the two are indistinguishable microscopically, the name of the more common mineral has been used in this paper.

---

[1] Bull. U. S. Geol. Survey No. 430, 1910, p. 88.

In specimens of alaskite porphyry wall rock the only alteration is the development of a small amount of chlorite and scattered crystals of pyrite. Tourmaline was noted in connection with quartz, sericite, calcite, and chlorite in the altered meta-andesite of the Mountain Monarch prospect.

The changes in the meta-andesite—the uralitization of the pyroxene and the development of secondary epidote, chlorite, calcite, and quartz—are not confined to the vicinity of the ore deposits and are products of regional rather than hydrothermal metamorphism. Compared with the hydrothermal alteration of the rocks of the Sierra Nevada camps as described by Lindgren,[1] there seems to be less alteration of the walls and a smaller development of the carbonates and potash-bearing micas in the altered porphyries.

MINING CONDITIONS.

The rich surface ores of the known veins have now been practically exhausted, but the persistence of ore at considerable depth has been shown in the development of the Gladstone and Mount Shasta veins. There is no reason to suppose that these are exceptional, and though there is an undoubted decrease in tenor below the zone of oxidation, it is believed that under careful management many of the veins could be worked at a profit to considerable depths in spite of the increase in cost. At the present time the Gladstone and the Mount Shasta are the only mines which are not developed almost entirely by tunnels.

The region is particularly favored in its natural features, as the rugged topography allows extensive development by tunnels. The ore is free milling and easily crushed, and water power is everywhere available. In the northern part of the area timber is abundant.

The peculiar pocket deposits of the area have given rise to a class of prospectors known as "pocket hunters." These men follow carefully the contacts of slate and meta-andesite and by systematic panning discover many rich pockets by tracing the particles of gold in the soil to their sources. As soon as a pocket is gouged out and the joint plane or contact where it was found no longer shows colors, the place is abandoned. Possibly future exploration of pocket zones along the contact to somewhat greater depths, particularly where there are calcite lenses, may reveal other deposits of the type of the Five Pines mine.

It is impossible to say whether new discoveries of the fissure-vein type may be expected. Areas in the vicinity of masses or dikes of soda granite porphyry in slate of the Bragdon formation should be prospected carefully.

---

[1] Lindgren, Waldemar, Characteristic features of California gold-quartz veins: Bull. Geol. Soc. America, vol. 6, 1875, pp. 221-240.

Many of the failures in lode mining have been due to the installation of mills more elaborate than the size of the ore body justified. The owners of some of the smaller veins which are worked profitably with a mill of two to five stamps would lose money by attempting to operate on a larger scale.

MINES.

WHISKEYTOWN (STELLA) DISTRICT.

The Whiskeytown or Stella district is probably second to French Gulch in total output, though none of the mines are at present large producers. The mines lie along the edge of a mass of alaskite porphyry which borders a larger area of quartz diorite and is intrusive into the meta-andesite. The slates and conglomerates of the Bragdon formation appear near the heads of Whiskey and Grizzly creeks. Numerous small dikes of quartz-augite diorite cut the alaskite porphyry and granodiorite, and two larger dikes of diorite porphyry cut the Bragdon formation in the northern part of the district.

MOUNT SHASTA MINE (1).[1]

The Mount Shasta mine (Mount Shasta Mining Co., owner; Guy M. Vail, manager) is in the quartz diorite mass which covers the southeast corner of the quadrangle and is about 3 miles south of Whiskeytown and a mile west of the eastern boundary of the quadrangle.

The deposit was discovered in 1897 by George Leversay, who, with his partners, took out 88 tons of oxidized ore that ran $48.44 to the ton, giving a production of $4,263. It was sold to the Mount Shasta Gold Mines Corporation, which continued development and mined altogether from the first six levels (398 feet) a total of 4,072 tons, averaging $42.69 a ton, or $173,876, giving a total production of about $178,000.[2] The old company failed in 1905, largely through unfortunate ventures in other directions, and the property remained idle until 1911, when the present company began development work on the seventh level (465 feet). In July, 1912, four men were employed.

The country rock of the region is quartz diorite, but the ore deposit itself lies within an elongate mass of alaskite porphyry about 300 feet wide. This rock is fine grained, dense, and aphanitic. The only phenocrysts are small quartzes, up to 2 millimeters in diameter, and rare feldspars. In places, especially near the vein, the rock is much sheared and has the appearance of a flow-banded rhyolite. Secondary quartz has been introduced in lenses along the shear planes.

---

[1] Numbers refer to map (Pl. II).
[2] Figures furnished by Mr. G. M. Vail.

Under the microscope the rock shows small phenocrysts of quartz and albite feldspar, irregular and broken but fairly fresh. The groundmass is a microcrystalline aggregate of quartz and feldspar, the former predominating. Small patches and shreds of sericite and chlorite indicate secondary hydrothermal action. Irregular but roughly parallel veinlets of quartz and sericite cut the rock.

Where seen adjacent to the vein (on the seventh level) the porphyry is in places partly silicified; elsewhere near the vein it is chloritized; in both situations a small amount of pyrite has been introduced. The zone of intense alteration of the porphyry does not extend more than 15 feet from the vein, and the silicified and chloritized porphyry carries no gold. In the siliceous phase the alteration consists in the introduction of quartz along microscopic but closely spaced fissures. Calcite appears in small specks, particularly near the feldspar phenocrysts, and in thread-like veinlets that cut the quartz. Where chloritization has been more prominent, chlorite, with a small amount of calcite, replaces the groundmass and a part of the feldspar phenocrysts. The quartz and feldspar phenocrysts also contain shreds of sericite.

About 200 yards northeast of the shaft, near the border of the alaskite porphyry, is a small outcrop of a very fine grained quartz-augite diorite which shows a larger proportion of ferromagnesian minerals as well as a finer grain than the typical diorite of this vicinity. Augite, almost entirely altered to chlorite, is the dominant mineral. Quartz and andesine feldspar are subordinate.

The two veins worked are parallel in strike and about 50 feet apart on the surface. Near the surface the dip of both veins is to the southwest. The dip of the east vein changes to northeast between the second and third levels, and that of the west vein between the fourth and fifth levels.

The ore is white quartz, as a rule so much sheared and fissured as to be very friable. In places, however, it is firm and massive and is frozen to the walls. Even where it is most completely shattered some large individual crystals an inch or more in length may be seen. A little calcite in small crystalline masses is scattered irregularly through the quartz. Near the walls and in the altered wall rock inclosed in the quartz are small patches of sericite. Rarely a little kaolin is also present.

Pyrite is the only metallic mineral of any abundance. It is found in bands of crystals in the quartz generally close to the walls and to a slight extent as an impregnation of the altered alaskite porphyry. In the shattered white quartz from the seventh level small specks of molybdenite were seen. This mineral has not been observed in the ore of the upper levels and so far as known does not occur elsewhere in the quadrangle.

Work at present is confined to drifting toward the north along the west vein in the expectation of encountering the ore shoot mined on the level above. The quartz in which the drift was being run in July, 1912, averaged about $4 a ton.

BLACKSTONE PROSPECT.

About 1,200 feet north of the Mount Shasta mine the Blackstone prospect, in quartz diorite, shows ore composed of auriferous pyrite in a gangue of quartz and dolomite cut by minute threads of specularite.

MOUNTAIN MONARCH PROSPECT (2).

The Mountain Monarch is a copper prospect about 2 miles due south of Whiskeytown, on the flat-topped ridge west of the valley of Clear Creek. The workings consist of a small shaft on the top of the ridge at an elevation of about 2,400 feet, filled with water at the time of visit, and a tunnel in the hill, about 400 feet below the shaft, which has been driven 720 feet of the 1,200 feet that it has been calculated is necessary in order to reach the ore body shown in the shaft.

The country rock at a distance from the ore is a much sheared and epidotized meta-andesite, in places slightly pyritized along the joint planes. The prospect is not far from the contact of the meta-andesite and quartz diorite, and small dikes of alaskite porphyry cut the meta-andesite in the vicinity of the tunnel, though none were seen in the tunnel itself.

A few tons of ore has been stacked near the shaft. A part of it is almost entirely pyrite, in crystalline masses, in which the crystals vary from minute specks to bodies about 3 millimeters in diameter, but here and there are small amounts of glassy quartz. Small velvety feather-like clusters of minute tourmaline needles are common in parts of the pyrite, especially where the grain is finest. Besides the pyrite the only other metallic minerals present are rare specks of chalcopyrite distributed irregularly throughout the ore and a coating of covellite over a part of the pyrite. Other specimens of the ore consist of meta-andesite, in part replaced by pyrite and accompanied by small clusters of tourmaline needles, chlorite, epidote, and a little quartz.

No data could be obtained as to the size or shape of the ore body or the value of the ore.

MASCOT MINE (3).

The Mascot mine (Gray & Rossi, owners) lies about 2½ miles southeast of Whiskeytown. The workings consist of two tunnels, on the upper of which a vein has been followed for about 200 feet. It is planned to crosscut the veins in the lower tunnel and to erect a 10-stamp mill. So far there has been no production.

The country rock is quartz diorite, here rather deficient in ferro-magnesian minerals, grading off within half a mile to the west and northeast into the coarser-grained type of alaskite porphyry. The diorite is cut by a dike of fine-grained quartz-augite diorite, 100 feet or more in width. It is similar in mineral composition to that of the Mount Shasta mine, except that the augite is less altered. A similar dike about 40 feet wide has been crossed in the lower tunnel.

The two veins lie entirely within the basic dike near the contact with the diorite. At one point there is also a little quartz on the contact of the dike, but it has not been worked. One vein strikes N. 55° E. and dips 43°–52° SE.; the other strikes N. 50°–62° E. and dips 57°–70° SE. Only the first has been developed. Its width varies from 6 inches to 3 feet. The quartz is in places much crushed, and there is always considerable gouge on the hanging wall and locally on the footwall as well. The quartz is as a rule distinctly crystalline, and small vugs are common. The ore is entirely oxidized with the exception of a few partly altered specks of pyrite. Generally it has a platy appearance, with layers of quartz 2 to 4 milli-meters thick separated by dark planes of manganese oxide. Manganese oxide is much more prominent than in other mines in the quadrangle and gives the quartz a dark-gray appearance, diversified here and there by minute spots of yellow iron oxide.

Two ore shoots have been prospected to some extent by raises. These are said to be each about 100 feet in length along the drift and to show a value of $11.85 a ton.

GAMBRINUS MINE (4).

The Gambrinus mine (Shasta Monarch Mining Co., owner; T. W. Rogers, superintendent), lies on the east bank of Whiskey Creek directly opposite Whiskeytown. The deposit was discovered about 1870 in the course of placer mining and has passed through several hands. The earlier work consisted of gophering along rich surface streaks. The total known production is $127,000, which is exclusive of an unknown amount obtained by "snipers" and lessees. The present company has been in possession for the last three years and up to July, 1912, had produced about $5,000.

The veins lie in alaskite porphyry near the contact with meta-andesite. On the opposite side of the stream and a few hundred feet to the north is an outcrop of a basic dike similar to that at the Mascot and Mount Shasta mine.

The alaskite porphyry is of the fine-grained type characteristic of the vicinity. Near the veins it is a blue-gray rock, much jointed and stained by iron oxide, carrying small phenocrysts of quartz and feldspar.

Four veins are exposed on the property. The development work consists of shallow surface workings and a crosscut from a few feet above the stream level, with drifts on the two southwestern veins and a lower level 40 feet below that was filled with water in July, 1912.

The veins are approximately parallel and have strikes varying from N. 50° W. to west and dips from 45° N. to vertical. The three northeastern veins all lie within a 60-foot zone. In these veins the quartz forms irregular lenses in zones of crushed alaskite porphyry, 3 or 4 feet wide. Small grains of albite were observed under the microscope in the quartz near the walls. These crushed zones are persistent and are drifted on when the quartz pinches out. The ore in the upper workings is oxidized but shows considerable pyrite in the pan, as well as free gold. The quartz of the northeastern vein carries no manganese. On the main level there is a small amount of pyrite together with free gold. Tellurides are reported, but their presence is doubtful. Small copper stains are seen here and there in the quartz. The alaskite between the veins is silicified and carries considerable pyrite. Assays of this altered rock for the 27 feet between two of these veins is reported to have shown a tenor of $8 a ton, practically all of which was in the pyrite. Under the microscope, the only apparent change aside from the introduction of pyrite is the development of chlorite and possibly a slight silicification of the groundmass. A few patches of coarsely crystalline pyrite occur in the porphyry. The crystals of pyrite may be as much as 6 or 8 millimeters across and are said to be barren.

The southwestern vein is not exposed on the surface and is more distinct and regular than the other three. It consists of 1½ feet of manganiferous quartz carrying free gold. No calcite was seen in any of the ore from this mine.

On the lower level, which was not accessible when the mine was visited, there have been found, according to Mr. Rogers, small amounts of chalcopyrite with which free gold is often associated.

A 10-stamp mill is now being set up on the property to replace the small prospecting mill formerly used.

MAD OX MINE (5).

The Mad Ox mine (Caribou Gold Mining & Power Co., Trinity Center, owner), on the south side of Mad Ox Gulch near its junction with Whiskey Creek, has not been in operation for the last two years. No data could be obtained as to former production or value of the ore.

The vein lies in a zone of intensely sheared meta-andesite 4 to 6 feet wide, near an intrusion of alaskite porphyry. The alaskite where close to the vein is silicified and in part replaced by calcite. The strike of the vein is N. 22°–33° E. and the dip between 80° SE. and vertical. In width the vein varies from a mere streak of gouge to a maximum of 4 feet. The longest portion of the vein that is wide enough to stope is about 100 feet long. The ore is somewhat iron stained quartz with a little calcite but carries no visible sulphides.

The Mad Mule mine (Mad Mule Gold Mining Co., of San Francisco, owner; T. W. Briggs, superintendent), formerly the Banghart, is one of the best examples of the pocket type of deposit found in the quadrangle. The mine occupies almost the whole length of Mad Mule Gulch, a tributary of Whiskey Creek about 3 miles north of Whiskeytown. The ground was located and the surface sluiced in the early fifties. At that time the length of claims allowed was only 200 feet, and as the deposit continues for nearly a mile the number of old workings is tremendous. Mr. Briggs estimates that the total production has been about a million dollars, exclusive of a large but unknown amount taken out at different times by "snipers."

The ore lies close to a diorite porphyry dike which cuts the meta-andesite, the alaskite, and the slate and conglomerate of the Bragdon formation.

The intrusive rock is extremely striking in appearance, large white feldspars, up to a centimeter in length, being thickly studded through a fine-grained gray groundmass containing minute hornblende needles. Away from the zone of weathering, which has a maximum depth of about 50 feet, the rock is generally fresh and hard. A little pyrite scattered in small cubes throughout the rock constitutes the only indication of hydrothermal alteration. Under the microscope the feldspar phenocrysts appear to be albite grading toward oligoclase. The hornblende is completely altered to chlorite, together with some secondary calcite and a little iron oxide. The groundmass is made up of minute feldspar microlites. Calcite also occurs in irregular patches and microscopic lenses and fissures. In many of these there is an outer border of quartz, suggesting that an open cavity had been first lined by quartz and later filled with calcite.

At the west end, where the workings have been most extensive, the dike is about 150 feet wide. Its strike is here generally west, though with many local variations, and its dip 40°–60° N. Farther downhill to the east, at an elevation of 1,900 feet, where the company is running its present tunnel, the width is only about 12 feet and the dip nearly vertical. The dike, though slightly faulted here and there, is comparatively free from the intense shearing that characterizes the slate, meta-andesite, and alaskite porphyry on its walls.

The alaskite porphyry carries prominent quartz phenocrysts and is intermediate in grain between the extremely aphanitic type found on South Fork Mountain and the more granitic phase exposed along the stage road east of Whiskeytown.

The Bragdon formation as represented in the vicinity of the mine consists chiefly of black carbonaceous slate, much sheared and similar to that found in connection with the ores of the French Gulch district. Near the surface iron staining is common, and a few small circular rosettes of gypsum needles are to be seen on the shearing planes of the slate.

The workings consist of a large number of tunnels along the walls of the dike. They are more numerous and, according to local report, find richer ore on the hanging wall. Along the contact and in the porphyry and slate to distances of a few feet from the contact are small quartz stringers 1 to 2 inches wide. The quartz is honeycombed with cavities, many of which are partly filled with sooty manganese oxide. Many of the cavities are drusy, with small quartz prisms, 2 or 3 millimeters long, growing out from the flat faces of larger crystals. Others are interstitial between larger quartz crystals, and in many of these minute manganese dendrites are arranged along two series of lines intersecting in rhombs, which suggest that the manganese was deposited along cleavage cracks in calcite and that the cavity was formed by the leaching out of calcite. Other cavities are in the form of small parallel gashes in the quartz. These are commonly irregular in detail, though in a few of them the outlines are so sharp and clear as to look like cuts made by a saw. These cavities also appear to have been formed by the leaching out of calcite. The slate is, moreover, cut by numerous small stringers of pyrite and calcite. According to Mr. Briggs, this pyrite is auriferous. No arsenopyrite was found, but some arsenic is said to be present in the concentrates.

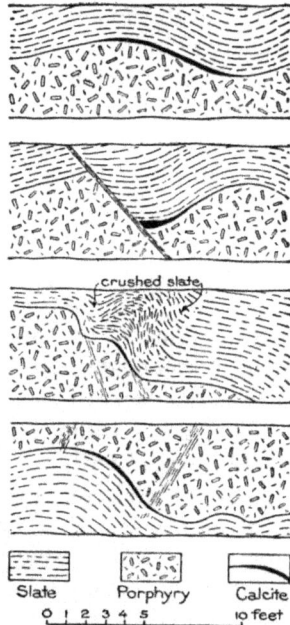

FIGURE 3.—Sketch showing position of pockets, Mad Mule mine, north of Whiskeytown, Cal.

So far as could be observed, no work has been done on veinlets of the type just described.

The ore mined is almost entirely calcite, which occurs at intervals along both hanging and foot wall in small lenses known as "points." (See fig. 3.) These points have a variety of forms. The common form is a flat lens of calcite generally not over 4 inches thick, deposited in the trough formed by some irregularity in the contact of the dikes and slates. Most commonly the calcite is not directly in the crotch of the trough but along one side, tapering out toward the crotch. Such "points" as have been mined on the footwall side of the dike seem to be most commonly in the reverse position—that is, to be formed near the top of an arch rather than in a trough. These calcite lenses are as a rule not more than 3 or 4 feet in length along the strike and have been followed upward for as much as 20 or 30 feet. The calcite is rarely drusy, but here and there small acute rhombohedral crystals project into open spaces. Very minor amounts of quartz as small patches are inclosed in the calcite and are apparently contemporaneous with it. In some specimens small specks of a green micaceous mineral were seen.

Pyrite is the only sulphide present, but it is comparatively rare in the calcite. It is found in small lines traversing the ore and in scattered cubes. In the troughs, all of which are present watercourses, it occurs near the ends of the calcite lenses, though within the slate, as small octahedra, unmodified by other forms.

Gold has been mined only in the "points." Most commonly it forms a thin film on the surface of the calcite at the junction with the slate, rarely extending to plates of noticeable thickness. It is said that the largest single piece of gold taken from the mine was in the form of a plate nearly a quarter of an inch thick and weighing over 100 ounces. Gold also occurs in much smaller masses entirely within the calcite, here following the cleavage planes. Invariably, however, it is close to the slate, being nowhere, as far as could be seen, over half an inch distant. The intersection of the manganese-bearing quartz stringers referred to above with the calcite of the "points" is regarded as an indication of a rich pocket.

The "points" mined have been chiefly on the hanging-wall side of the dike and all within the upper 600 feet. Moreover, all the "points" have been found comparatively near the surface. Few of the tunnels are more than 200 feet in length, and in the longest, about 1,000 feet long, pockets had not been mined beyond the first 500 feet, which would be equivalent to not more than 250 feet in depth below the surface. All these facts indicate that the gold content of the pockets is related to the present surface. Such an origin is likewise indicated by the position of the pockets which are closely connected with the present watercourses.

TRUSCOTT MINE (7).

The Truscott mine (John Martin, owner), formerly the Emigrant, is situated near the head of Grizzly Gulch, about 2 miles northeast of the Tower House and a mile west of the head of Mad Mule Gulch. The mine was discovered about 25 years ago and has produced about $60,000, of which $12,000 was taken out in the last three years. The veins have been developed by several tunnels at two levels 40 feet apart.

The principal vein consists of lenses of quartz along the footwall contact of a dike of andesite porphyry with black slate. The average strike of the contact is N. 20° E. and the dip 60°–80° W. The width of the dike is about 300 feet. Some ore has also been found on the hanging-wall side, but none of it has been developed.

The ore occurs in large lenses along the contact. The only developed lens shows a length of about 100 feet and a greatest width of 10 feet. At the end the lens tapers down to a mere streak of black gouge along the contact. This was followed for about 30 feet and a second lens encountered. All the quartz is said to be workable and to carry from $10 to $15 a ton in free gold. The gangue consists of white quartz with a small amount of calcite, much mixed with black slate, but without porphyry. The only metallic mineral observed in the ore was pyrite, though small copper stains were seen and chalcopyrite is said to be present in small amounts. The pyrite occurs in minute crystals in the quartz and the slate, but not, so far as could be observed, in the calcite. Branching out from the main vein into the porphyry are numerous small stringers of manganiferous quartz.

On the upper level a small vein 4 to 8 inches wide branches off into the porphyry at a small angle. Here the ore is iron-stained quartz much mixed with altered porphyry and carrying much visible gold in association with partly oxidized pyrite. The ore from this small vein is said to pan between $100 and $300 a ton.

BRIGHT STAR MINE.

A short distance northeast of the Truscott is the Bright Star mine, which was worked to a small extent during a local boom a few years ago but is now abandoned.

FRENCH GULCH DISTRICT.

PRODUCTION AND GENERAL FEATURES.

The French Gulch district is the oldest lode-mining district in the quadrangle, the Washington, the oldest mine in the region, having been located in 1852. The following fragmentary records of production give some idea of the importance of the mines in this district:

Production of French Gulch district.

	Gold.	Silver (ounces).
Year ending September, 1854 [a]	$53,232	....
1855 (8 months) [a]	22,132	....
Year ending July 1, 1869 [b]	50,044	....
Year ending July 1, 1871 [c]	44,639	....
Year ending May 31, 1881 [d]	38,500	....
1909 [e]	418,622	1,709
1910 [e]	560,144	4,683
1911 [e]	420,451	3,923

[a] Production of the Washington mine. Trask, J. B., Report on the geology of northern and southern California; Rept. California State Geologist, 1856.
[b] Chiefly Washington mine. Raymond, R. W., Statistics of mines and mining in the States and Territories west of the Rocky Mountains, 1870.
[c] Idem, 1872.
[d] Second Rept. California State Mineralogist, 1882. Apparently includes placer production.
[e] Mineral Resources of the United States.

The mines of the district lie within the area of slate and conglomerate of the Bragdon formation, which begins at Tower House and extends northward to the border of the quadrangle. With the exception of the Eldorado, which is more closely connected with the pocket type, all the deposits are simple fissure veins and show a close mineralogic similarity. With one exception all these veins are closely associated with intrusive soda granite porphyry or diorite porphyry, chiefly the former.

The principal fissures trend nearly east and west, though a few have a northerly strike.

ELDORADO MINE (8).

The Eldorado mine (J. G. Connors, owner; Garvin & Gatney, lessees), is on the west side of Mill Creek about half a mile south of the Tower House. The property was located by William Paul about 1885, and the total production has been estimated at $25,000, of which $3,500 has been obtained by the present lessees between August, 1911, and July, 1912.

The lode lies along the fault contact of meta-andesite and slate, which strikes about N. 20° W. and dips 50°–70° E. The only working now accessible is a 400-foot drift at an elevation of 1,250 feet. There is much gouge along the contact and a part of the faulting has been oblique to the original contact, resulting in sharp wedges of slate that enter the meta-andesite at small angles and small lenses of much-crushed slate that are included in the meta-andesite at 2 feet or less from the contact.

Quartz occurs in small lenses about 2 feet in width along the slate walls and in the meta-andesite, especially near the slate. The quartz is somewhat drusy, though the vugs are all small, never more than 2 centimeters in greatest diameter. The vein quartz is much mixed with fragments of meta-andesite, whose uncertain boundaries imply some replacement. A little sericite accompanies the altered meta-andesite in places. No calcite was seen in any of the ore.

On this level the ore is almost completely oxidized. Sooty manganese oxide partly fills many of the druses. Blotches of iron oxide show the alteration of pyrite. On the lower level, 90 feet below, pyrite was found in the quartz, but is said to have proved barren, all the values being in free gold.

The gold occurs entirely in pockets in the quartz and gouge. The richest pocket found, from which $2,500 worth of gold was taken, was entirely in the gouge between quartz and meta-andesite. Other rich pockets are in the form of quartz veinlets, generally less than an inch wide, in the meta-andesite near the infaulted lenses of slate. These veinlets carry visible gold in association with specks of iron oxide and in small flakes lining minute druses. So far as could be observed, the rich quartz was free from manganese.

The gold is noticeably light colored and its fineness is much below the average for the district. It is said that its average value is about $14 an ounce and that some of it falls as low as $12. No intrusive rock has been encountered in the workings, so far as known, and none was seen on the surface in the vicinity, though the ridge was not examined in detail.

GLADSTONE MINE (18).

The Gladstone mine (Hazel Gold Mining Co., owner; J. O. Jillson, managing director; E. Young, superintendent) lies on the north side of Cline Gulch, about 5 miles by road from the town of French Gulch. It was originally located in 1896 by T. Cumming. After the oxidized ore was exhausted the mine changed ownership more than once and was purchased by the present company in 1901. The production previous to 1901 was about $85,000. From February, 1901, to June, 1912, the production has been, in gold bar, $2,389,491.78; in concentrates (net), $109,739.90; total, $2,499,231.68. Under present conditions the annual production is about $360,000.

The steep topography of the region has made it possible to mine the upper 1,000 feet of the vein by means of tunnels. The lower portion is worked from a blind shaft on the main adit (Ohio) level, which in October, 1912, had reached a depth of 1,080 feet. This part of the vein is developed by three levels. The exhausted stopes and upper workings are filled with the waste obtained in crosscutting.

The company owns a power plant on Crystal Creek which furnishes power for the mill, electric haulage, and outside lighting. The power for the hoist and pump is supplied by the Northern California Power Co. There is a 30-stamp mill, four Wilfley tables, and four vanners, the whole plant having a capacity of about 100 tons a day.

The country rock of the vein consists of slate, sandstone, and conglomerate of the Bragdon formation. The slates show great contortion, with much minor faulting and variations of dip. Conglomerate and sandstone are present only in minor amounts.

The vein itself differs from most deposits of this type in that it does not cut any igneous rocks. On the adit level several hundred feet from the vein is a small dike of much altered diabase. The dike is about 2 feet wide and much broken by the numerous small faults that cut the slate but do not affect the ore. On the fourth, fifth, and sixth levels, which were not accessible at the time of the writer's visits, the crosscuts from the shaft to the vein encountered a soda granite porphyry somewhat similar to that seen at the Milkmaid mine and elsewhere, though more granitic in texture. According to Mr. Young, this rock occurs in irregular masses about 150 feet south of the vein. These vary from a few feet to more than 40 feet in length and occur nearly vertically one over another. There is a small amount of gouge at the contact with the slate. No porphyry was found below the sixth level. Apparently these masses represent irregular apophyses from some large mass below. The most numerous phenocrysts are biotites from 1 to 3 millimeters in diameter in rough parallel arrangement. · There are also a considerable number of kaolinized oligoclase feldspar crystals from 2 to 4 millimeters in length. Quartz is in clear distinct grains, many of which show embayments due to magmatic corrosion. The groundmass consists of minute crystals of quartz and feldspar. It also contains a few grains of a colorless mineral, probably garnet. The alteration shown is entirely such as could be accomplished by surface waters, and no minerals characteristic of hydrothermal alteration are to be seen. In fact, there seems to be little or no hydrothermal alteration of any of the wall rocks as a result of the introduction of the vein-forming solutions, though in places thin bands of sandstone are impregnated by pyrite and rarely the slate near the vein is silicified.

The vein is inclosed in a zone of much crushed slate and sandstone 60 or 70 feet in width, but within this zone the vein itself is irregular. Where the walls are slate it tends to break up into small anastomosing veinlets, which must be worked as a whole, involving the mining of much waste material. In the conglomerate and sandstone the vein is much more distinctly defined, though narrower. In the crushed slate the greater proportion of quartz is on the footwall side of the crushed zone. Gouge is nearly always present on both walls. The common width of the ore as stoped is $2\frac{1}{2}$ feet, though in a few places it reaches a width of 12 feet. Along the adit level the vein has been followed for a distance of 2,000 feet. The present depth of working is about 1,000 feet below the adit level and 2,000 feet below the outcrop.

The strike is east with small local variations. From the summit of the outcrop to the adit level, 1,000 feet below, the vein is vertical. From the adit to the seventh level the dip is over 60° S., but below the seventh level the dip changes to steep north. This change in dip, together with the fact that the vein is much tighter and no longer carries the large amount of gouge that characterizes it on the upper levels, has led to the belief that the vein splits between the sixth and seventh levels and that it is the northern branch which has been followed. It is planned to test this hypothesis by crosscutting to the south from the ninth level.

The ore is, as a rule, massive white quartz, showing in places minute fissuring but very rarely any tendency toward crystal forms. According to Mr. Young, the quartz that occurs within conglomerate or sandstone walls tends to be more crystalline and vuggy. Such vugs as were seen were very small, not over a centimeter in length, and the individual quartz crystals are all under 3 millimeters in length. Most specimens of the vein quartz have a gray to bluish color due to the intimate mixture of fragments of slate wall rock with the quartz. A few small gray bands in the quartz appear to be due to shearing and consequent smearing out of slate fragments in lines parallel to the walls. Calcite is present in small amount throughout the ore. It is generally in small specks not over 1 or 2 centimeters in diameter but more rarely forms distinct lenses several inches long. Sericite is developed to a small degree in the replaced fragments of slate in the quartz.

The metallic minerals of the ore are pyrite, free gold, galena, sphalerite, and arsenopyrite. Pyrite is by far the most common of these. In the quartz itself it occurs in minute crystals less than a millimeter in size and almost invariably in close association either with the included fragments of slate or with the slate of the wall rock. In thin bands of sandstone in the wall rock pyrite is commonly present in small amount, scattered through the rock in minute crystals. In the carbonaceous slate, however, the pyrite does not occur as an impregnation but most commonly in small veinlets generally less than 5 millimeters in width, associated with a small amount of quartz. So far as could be observed, pyrite is not present in the calcite. The other sulphides, galena, sphalerite, and arsenopyrite, are extremely rare and are not found outside of the vein itself. Galena is present in small cubes, almost always close to the walls. The largest seen was about 2 millimeters. Sphalerite occurs in very minute specks and is much less common than galena. The presence of arsenopyrite is indicated by a small percentage of arsenic in the smelter returns of the concentrates, and the small white metallic specks seen here and there in the ore are shown to be this mineral when examined microscopically. The total of the concentrates forms less than 1 per cent by weight of the ore milled, although about 6 per cent of its value. Visible gold is of less common occurrence than in the other mines of the region, but many small specks are seen in the quartz near the included fragments of slate or the slate walls. About 94 per cent of the value of the ore is in free gold, recovered on the plates of the mill.

In the upper workings three separate ore shoots were mined. Below the adit level the two western shoots had joined and between the sixth and seventh levels the third ore shoot joined the others (fig. 4). The maximum drift width of any ore shoot was about 500 feet near the junction of the central and westernmost shoots, and the minimum slightly less than 200 feet.

In the oxidized zone, which, according to Mr. Young, reached a depth of only about 75 feet, scattered patches of very rich ore were

encountered. Below this zone, however, the tenor of the ore has
been rather even, the only noticeable change being a slight increase
in the amount of arsenopyrite. The ore as milled runs about $10 a
ton, but this is lower than the value of the quartz, as a large amount
of waste must always be milled and everything which when panned
shows an estimated tenor of $3 or over is mined. The good ore of
the ore shoots carries from $30 to $50 a ton, and some small stringers
and patches may run up into the hundreds of dollars.

This mine has been developed to a greater depth than any other
in the Weaverville area, and the fact that ore has been followed for

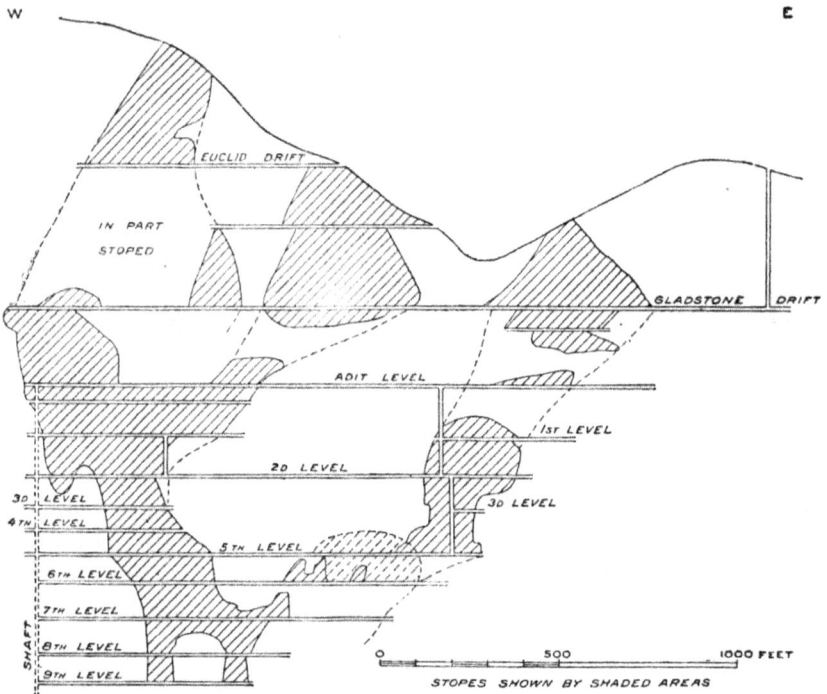

FIGURE 4.—Section along Gladstone vein, near French Gulch, Cal., showing position of ore shoots.

a vertical distance of 2,000 feet should be a good indication for the
persistence in depth of other deposits of this type.

AMERICAN MINE (19).

The American mine was located in 1887. The surface ores were
very rich, and it is reported that $2,000 was obtained from ore
crushed in a hand mortar.[1] The workings occupy the ridge north of
Cline Gulch between Clear Creek and J-I-C Gulch and consist of
tunnels at several elevations between 2,500 and 3,150 feet. The

_____
[1] Eighth Ann. Rept. California State Mineralogist, 1888.

mine has not been in operation for several years and many of the tunnels are caved. The production is unknown.

The country rock is slate and conglomerate in the upper workings and entirely slate below. Near the vein it is much fissured and disturbed but maintains a general dip to the south at varying angles. No igneous rock was seen either in the workings or on the dumps.

The vein lies in a narrow zone of much-crushed slate and varies in width from a feather edge to a maximum of about 3 feet. It has been stoped only in the eastern portion. The strike is from N. 80° E. to east and the dip from vertical to 80° S.

The unoxidized ore, so far as could be seen from the specimens picked from the ore bin, closely resembles that of the Gladstone mine in general appearance. Like the Gladstone ore, much of it has a bluish-gray color, due to specks of carbonaceous matter from the slate, and contains many small angular fragments of slate surrounded and partly replaced by quartz. The quartz contains numerous vugs and shows a crystalline form more commonly than that of the Gladstone ore. Most of the crystals appear to have grown outward from the slate fragments included in the vein. No calcite was seen in any of the specimens collected.

The metallic minerals appear to be rather less in quantity than in the Gladstone ore. Arsenopyrite, pyrite, and gold were the only metallic minerals seen. All of them were found close to the wall or near included fragments of slate. Arsenopyrite is the most common and is found in fair-sized crystals (2 millimeters or less) in the quartz and at the contact with the quartz and slate, both the walls and the included fragments. Pyrite is not common and occurs only close to the wall and in the slate. Gold was seen in small plates and imperfect crystals close to the contact of quartz and slate and in the dark bands in the quartz.

FRANKLIN MINE (20).

The Franklin mine (Western Exploration Co., owner; H. F. Musser, manager) is on the north side of French Gulch, about 3 miles northwest of the town. It is developed by adit levels at elevations of 1,950 and 2,055 feet, connected by a raise on the vein, and by a lower level 130 feet below the main level, reached by a winze. Drifts have been run on the vein in all three levels. As far as possible the stopes have been filled with waste. The large proportion of waste which it is necessary to mine brings down the value of the ore milled to about one-third of the assay value of the quartz. The ore is crushed in a 10-stamp mill on the neighboring Milkmaid property, and the concentrates are saved on four vanners. The mine is said to have been located in 1852, and is therefore one of the oldest in the State. The earlier production is unknown. The production for the last five years, according to Mr. Musser, was $350,000. In July, 1912, a force of 14 men was employed, and two lessees were working on the east end of the vein on the lower level.

The vein cuts both the slate and a mass of intrusive soda granite porphyry. A few hundred feet to the north the slate overlies a mass of meta-andesite. Dikes of diorite porphyry and quartz diorite porphyry outcrop a short distance to the south. The slate is here much sheared and crushed, having a glistening appearance due to the close slickensiding.

The intrusion of soda granite porphyry is irregular in form, but appears to occupy the valley of French Gulch at this point and to extend for about 1,500 feet northward. As seen in the mine workings the northern contact dips to the north, the upper workings being in slate and the lower in soda granite porphyry. The slate, except for the crushing and shearing, does not show any alteration near the vein. The porphyry, on the other hand, is altered and impregnated by arsenopyrite and pyrite. Under the microscope this change is seen to consist of the sericitization of the feldspar, with some development of secondary calcite, and the complete alteration of the biotite to a mixture of chlorite and sericite with specks of epidote. Sericite shreds also occur in the groundmass, which may be somewhat silicified as well.

Two veins have been followed to some extent in the Franklin workings; one with a strike of N. 5°–30° W., and a steep dip to the east, the other with a strike approximately west and a dip about 70° N. The intersection of the two veins has not been discovered. The northward-striking vein is in the extreme western part of the workings and has been stoped on the main level for about 70 feet only. A short distance above the level it passes out of the porphyry and into the slate, where it splits up into a series of small stringers. The other vein has been followed in all for 700 feet on the main level, where the walls are porphyry, and for smaller distances on the upper and lower workings. In the upper level (105 feet above the main level) the walls are slate and the vein less defined than in the dacite porphyry. Besides the two main veins small stringers have been gophered for short distances. The maximum width of the vein is 4½ feet, but in places it narrows to a few inches. Above the upper level there are old workings on the hillside, now inaccessible, from which, it is said, extremely rich oxidized ore was taken.

The gangue minerals are quartz and calcite. Much of the quartz presents a faintly mottled grayish appearance, probably due to the inclusion of minute particles of carbonaceous matter from the slate wall rock. Fragments of almost completely replaced dacite porphyry are likewise common. Calcite is less abundant than in the Gladstone ore but is present in small patches in the quartz. The proportion of calcite in the ore appears to be increasing in depth.

The sulphides in the ore, in the order of their prominence, are arsenopyrite, pyrite, galena, and sphalerite. Arsenopyrite occurs in small well-formed crystals rarely over 3 millimeters in length. Its characteristic position is close to the porphyry walls or in scattered crystals in the altered porphyry close to the vein. Arsenopyrite also occurs in small quartz stringers in the porphyry, but was not seen in the larger veins. The pyrite is also found with the arsenopyrite as an impregnation of the porphyry, but is more prominent where slate forms the wall rock. Although more prevalent in the wall rock than in the quartz, it is more general in its distribution than arsenopyrite. In the vein itself the pyrite is usually present as small specks and crystals, closely associated with sphalerite and to a less extent with galena. Galena and sphalerite are entirely confined to the vein. Galena is the more abundant of the two minerals, and though its most common position is near the walls or included fragments of slate it also occurs in small irregular patches a few millimeters in diameter in the quartz itself, entirely without relation to the walls. Sphalerite is similar to galena in its distribution. Neither mineral shows distinct crystal outlines. Altogether the sulphides, according to Mr. Musser, the superintendent, amount to about 0.75 per cent of the weight of the ore and carry $150 to the ton in gold.

Visible gold is more abundant in this ore than in the average ore of this type. In all specimens examined, in which the gold is present in sufficient size to be visible with a hand lens, it is in close association with the galena.

The oxidized ores were all mined in the early days, and no data could be obtained as to their value or the depth to which oxidation extends.

The ore of the main vein, which runs at its best about $45 a ton, is in rather irregular pay shoots that pitch steeply to the south. The best ore is commonly found where the vein is in the slate close to the porphyry contact. In the lowest level, 130 feet below the adit and about 200 feet below the point where the vein crosses the slate and porphyry contact, the ore is said to decrease in value, although it presents the same general appearance.

MILKMAID MINE (21).

The Milkmaid mine adjoins the Franklin on the east and is under the same ownership. No work has been done for several years, and the production is unknown.

The rocks in the vicinity consist of soda granite porphyry similar to that at the Franklin mine, diorite porphyry, and much-sheared slate of the Bragdon formation. Both porphyries are intrusive into the slate, but their relative ages could not be determined.   The soda granite porphyry is a part of the mass exposed at the Franklin.   The diorite porphyry is a dark-gray aphanitic rock.   The only minerals visible under the hand lens are phenocrysts of white feldspar, some of them as much as 5 millimeters square, which are thickly scattered over the rock.

In one tunnel a vertical vein striking N. 10°–30° E. near the contact of the slate and diorite porphyry had been stoped to some extent. What ore was seen consists of quartz intergrown with a little calcite and rare pyrite.   To the south is an incline shaft apparently following another vein.   The dip of the incline is 44° and the direction N. 60° E. The quartz on the dump is very glassy in appearance, shows vugs with crystals two centimeters in length, and is iron stained.

WASHINGTON MINE (22).

The Washington mine (Washington Gold Mining Co., owner; Maxwell & Ketch, lessees), covers the greater part of the hill between the two forks of French Gulch.   The mine was located in 1852 and is said to be the oldest in the county and one of the oldest in the State. From September, 1853, to September, 1854, its output was $53,232. At the close of 1855 the workings consisted of three levels of 522, 222, and 97 feet and three shafts of 34, 12, and 23 fathoms.[1]   In 1869 there was a 22-stamp mill on the property and the production was $45,722.[2]   In 1871 a production of $31,153 was reported.[3]   In 1890 the total production up to that date was reported as between $500,000 and $600,000.[4]   The exact total is unknown, but persons familiar with the district estimate it at one to two million dollars.

No regular work has been done for several years, but lessees have been at work more or less regularly in different parts of the property. The equipment consists of a 10-stamp mill run by water power and two vanners.   There are numerous tunnels, many of which are now inaccessible, connected by raises.   The workings honeycomb the hill from the creek level at about 2,000 feet to a point near the summit, at about 2,900 feet.

The lower workings are in the meta-andesite, which is overlain by slate.   These rocks are cut by a great number of intrusions of soda granite porphyry and diorite porphyry, which vary from sheets a few inches in thickness to dikes and irregular masses up to 200 feet or more thick.   The meta-andesite extends up as high as the sublevel 150 feet above the lower main tunnel elevation (2,400 feet).   Owing

[1] Trask, J. B., Report on the geology of northern and southern California: Rept. California State Geologist, 1856.

[2] Raymond, R. W., Statistics of mines and mining in the States and Territories west of the Rocky Mountains, 1870.

[3] Idem, 1872.

[4] Tenth Ann. Rept. California State Mineralogist, 1890.

to movement at small angles to the contact, irregular wedges of one rock penetrate the other and give the meta-andesite the appearance of an intrusion into the slate.

The intrusive porphyries in the vicinity of the veins consist of a biotite-bearing diorite porphyry that outcrops in two large sills in the southern fork of French Gulch, soda granite porphyry that forms a large mass to the south of the main (2,500-foot) tunnel, numerous small sills and dikes on the point of the ridge to the west, and diorite porphyry that outcrops near the mill about 500 feet south of the workings. The biotite-bearing diorite porphyry resembles the soda granite porphyry so common in the region, except for the lack of quartz and the more calcic nature of the feldspars. The soda granite porphyry is in every respect closely similar to that exposed in the Franklin workings and is probably a part of the same mass. The diorite porphyry shows numerous white-zoned feldspars (as much as a centimeter in length) in a dark groundmass.

Two veins have been worked on the property. One of these strikes about north and dips 60°–70° E. and appears to cut off the other, which strikes a few degrees north of east and dips at varying angles to the north. The north-south vein is a well-marked fault plane with a large amount of gouge, in which occur lenses of quartz having a maximum width of about 10 feet. The other vein pinches and swells to a considerable degree, but the ore as mined in few places was over 4 feet wide. Small stringers and veinlets branching off in every direction, have been followed in the hope of encountering rich pockets.

The ore is quartz with a small amount of sulphides. The two veins are similar, but the ore of the north-south vein may contain slightly more blende and less galena than the east-west vein. Quartz is the most abundant gangue mineral. It is white and massive, having in places a somewhat glassy appearance, and is comparatively free from the included carbonaceous matter which gives the gray color to the quartz of neighboring mines. Calcite occurs only in small specks in the quartz, and sericite was seen only under the microscope, in minute quartz veinlets which cut the pyrite.

The metallic minerals include pyrite, galena, blende, and arsenopyrite. Pyrite is by far the most prominent and is not, as in the Franklin ore, associated principally with the wall rock but is distributed throughout the ore, generally in small and rather distinct fissures in the quartz, usually not over 3 centimeters in length and 3 or 4 millimeters in width, and associated with small amounts of galena and in places with a little blende. Pyrite also occurs in small patches or individual crystals, the latter generally less than a millimeter square, and to some extent as an impregnation of the wall rock. Galena is found in small fissures associated with a larger amount of pyrite and to a less extent in small specks, about a millimeter wide, in the quartz, principally near the walls. Sphalerite follows the galena very closely but is far less in amount. Arsenopyrite was seen at only one point where soda granite porphyry had been completely silicified and impregnated with pyrite and arsenopyrite. The arsenopyrite is in very minute (0.5 millimeter or less), well-formed crystals, for the most part grouped close to fairly well-marked planes in the altered rock. Small crystals were also seen in small quartz veinlets which cut the silicified porphyry. Visible gold in the primary ore is found only in close association with the galena, and the presence of galena is looked upon as a sign of valuable ore.

The oxidized ore that was first mined was exceedingly rich and ran as high as $600 a ton. The first mining work done consisted in sluicing the rich and decomposed material on the outcrop. At present two lessees are working stringers of oxidized and partly oxidized ore on the upper part of the hill. These veinlets occur in close connection with the numerous small dikes of soda granite porphyry, and contain in places small pockets of very rich ore Where completely oxidized the ore is a white quartz with irregular reddish stains of iron oxide. Small honeycomb-like cavities 1 or 2 centimeters in length are scattered throughout the rock, representing the oxidation of the sulphides. All these cavities contain more or less iron oxide and are irregularly flecked with small leaf-like plates of gold, at the largest 3 or 4 millimeters in length.

Of the two principal veins the north-south vein was chiefly worked near the surface but was not found profitable below an elevation of 2,500 feet, and the east-west vein was principally worked below this level. On the latter vein the ore shoot appears to pitch west.

NIAGARA MINE (23).

The Niagara or Black Tom mine occupies a part of the same hill as the Washington but lies farther west. The property was located in 1857 but has not been worked for some years. The total production is estimated at somewhat under a million dollars.

The workings consist of several tunnels on both the north and south sides of the ridge. On the dumps, besides slate and conglomerate, are fragments of both diorite and soda granite porphyry, the latter in part more or less completely silicified, leaving phenocrysts of glassy quartz in a chalcedonic groundmass. So far as can be judged from the accessible lower workings, diorite porphyry forms the larger portion of the mass. Fine-grained quartz-augite diorite outcrops on the road near the lower tunnel, and a few pieces were found on the lowest of the several tunnel dumps.

Such ore as was seen was quartz, without any calcite. The usual sulphides are present in small amounts. Pyrite is most widely distributed, as it is found in the quartz, in the small stringers in the slate, and disseminated in minute crystals in the porphyry. Most commonly, however, it is in close association with the galena, here and there completely surrounding small fragments of that mineral. Arsenopyrite is in places in small crystals in the vein but is more commonly scattered through the altered porphyry in the wall. Galena and sphalerite are most abundant near the slate wall rock and include fragments of slate.

SUMMIT MINE (24).

The Summit mine (Joseph Porter and the Wheeler estate, owners; Allen & Alexson, lessees) is near the top of a spur running eastward from the county divide, about midway between the Brunswick and Niagara mines. Since 1907 the lessees have taken out about $30,000. The estimated total production is about $200,000.

A dike of soda granite porphyry from 25 to 40 feet wide cuts the slates and is itself crossed by two veins. The porphyry is rather finer grained than other rocks of this type and shows prominent biotite and feldspar phenocrysts, 1 to 2 millimeters in diameter, but no quartz.

The veins have been developed by three adit levels at vertical intervals of about 40 feet. The ore is quartz with a very subordinate amount of calcite and, near the walls, specks of metallic sulphides, pyrite, galena, sphalerite, and arsenopyrite, together with gold. For the most part the quartz is similar to that of other veins in the vicinity, massive and in part of a grayish tinge, but in places containing vugs 6 or 8 inches across. The oxidized ore shows staining by manganese as well as iron oxide. Of the sulphides, sphalerite is perhaps the most common and occurs in small crystals with galena and pyrite in the quartz at distances of less than a quarter of an inch from the wall. Arsenopyrite is in its usual position in the porphyry, close to the quartz, and in one specimen occurs in small streaks in the quartz near the wall. Visible gold in the vein itself was seen only in association with the arsenopyrite rather than around its usual nucleus of galena. In the large vugs gold is seen on many of the quartz crystals themselves, and in one specimen small flakes are inclosed in a large crystal, though possibly along a small fissure in the crystal.

Although one of the veins has been followed for as much as 300 feet, very little good ore has been encountered outside of the porphyry dike. Here ore running as high as $150 a ton is found in small ore shoots 20 feet or less in length along the drift, although on the second level the western vein showed workable ore all the way across the dike. Outside of the dike only a few small ore shoots have been worked. The richest ore mined, according to Mr. Alexson, was one 3-ton lot that milled $423 a ton.

The ore must be carted from the mine to the Washington mill on French Gulch, and hence mining is comparatively expensive.

The Brunswick mine (Brunswick Mining Co., of French Gulch, owner; H. D. Lacey, manager) is 2 miles south of the group of mines in French Gulch, on a steep ridge between Sawpit Gulch and Summit Gulch. The elevation of the tunnel is slightly less than 4,000 feet.

The mine was first located in 1879 and has been under its present ownership since 1906. The total production has been about $70,000, of which $45,000 was produced since 1906. There is a 10-stamp mill on the property, but it is not in operation, as the company is confining its attention to development work. In July, 1912, two men were employed.

The accessible workings consist of a tunnel that cuts entirely through the ridge and a drift along the vein. It is planned to prospect the vein at depth by a crosscut from the north side of the ridge at an elevation of 3,300 feet. The Bragdon formation is here less distorted than is common in the neighborhood of the veins and consists largely of black slate with a few conglomerate beds.

The only intrusive is a dike of augite-bearing diorite porphyry over 100 feet in width which strikes nearly east and west, though with a rather irregular contact, and dips about 60° N. The rock resembles that exposed on the Washington and Milkmaid properties, consisting of large white feldspar crystals in a gray groundmass, but the groundmass shows many minute augite prisms.

The ore lies entirely within the porphyry at a distance of 1 to 10 feet from the northern contact and consists of small lenses of quartz and rarely calcite in a narrow crushed zone. To a large extent the ore is oxidized and the white glassy quartz shows honeycombed cavities formed by the leaching out of pyrite and possibly also of calcite. These cavities, and the quartz as well, are deeply stained with iron oxide and also contain small black specks and streaks of manganese oxide. The only metallic mineral present is pyrite.

Not enough work has yet been done to indicate the size or inclination of the ore shoot. According to Mr. Lacey, the best ore so far found runs about $10 a ton.

The Accident or Sybil mine was idle in July, 1912. It lies at the north edge of the complex of dikes which marks the position of the group of mines near the head of French Gulch. Soda granite porphyry similar to that of the Franklin and diorite porphyry similar to the Brunswick dike cut the slate. The ore for the most part lies in the diorite porphyry near the contact of the slate, but in places the workings follow the contact itself, which here strikes about N. 80° W. and dips about 50° N.

The ore consists of blue-gray quartz with patches and streaks of white calcite, the latter sometimes as much as half an inch in width. The quartz has a mottled gray color, but the calcite is milky white and cut by veinlets of quartz and arsenopyrite 1 millimeter wide. There is the usual association of metallic minerals—arsenopyrite, pyrite, galena, and blende. Arsenopyrite is probably the most common. Its characteristic positions are as a band about a millimeter wide of very minute crystals bordering the calcite and in still smaller veinlets cutting the calcite, and in larger crystals (3 millimeters or less) scattered through the altered porphyry wall rock. Rarely small crystals appear in the quartz. Pyrite occurs commonly in irregular patches of crystals in the quartz, and to a less extent is present in millimeter-sized crystals in the altered porphyry near the vein. Galena is scattered through the quartz in patches and is commonest near the walls. Sphalerite is usually close to galena, but is less in amount and tends to be more generally scattered through the vein than the galena. Gold, wherever in plates large enough to be visible, is always close to the small patches of galena. Except for the small quartz and arsenopyrite fissures traversing the calcite, the sulphides occur only in the quartz or wall rock.

THREE SISTERS MINE.

The Three Sisters mine has evidently been abandoned for several years. The vein worked must be near a contact of soda granite porphyry with the slate, as both rocks are present on the dump. The ore is a white quartz with small veinlets of calcite that have weathered to a dark-brown color, implying the presence of manganese. Arsenopyrite and pyrite are the two metallic minerals present. The arsenopyrite is most closely associated with the altered porphyry, and the pyrite with the slate.

HIGHLAND MINE.

The Highland mine, at the head of Dutch Gulch, has a recorded production of $4,322 for 1869 [1] and $9,650 for 1871.[2] It has been idle for years, however, and was not visited.

SHIRTTAIL MINE.

The Shirttail mine, in the meta-andesite area on Drunken Gulch, was idle in July, 1912, and was not visited.

DEADWOOD DISTRICT.

LOCATION AND PRODUCTION.

The Deadwood district is separated from the French Gulch district, to the east, only by the county line. For all practical purposes it is a part of the same district. The geologic conditions are identical.

[1] Raymond, R. W., Statistics of mines and mining in the States and Territories west of the Rocky Mountains, 1870.
[2] Idem, 1872.

Dikes of porphyry of varying texture and composition cut the slates of the Bragdon formation and the veins are found in close proximity to the intrusive rocks.

The production of the district for 1909 [1] was 2,415 tons, valued at $31,094, and for 1910 [2] 2,723 tons, valued at $49,158. This gives an average tenor of $17 to $18 a ton.

Only one of the several mines of the district, the Brown Bear, was visited. Others are the Blue Jay, Lappin, Vermont, and Goodyear & Richards.

BROWN BEAR MINE (27).

The Brown Bear mine (Thomas McDonald, part owner; Barney McDonald, superintendent) is in the village of Deadwood, on the north side of Deadwood Gulch. The property was discovered in 1875 and has been worked ever since. The present Brown Bear property is a consolidation of several mines and taken as whole it has undoubtedly been the largest producer in the quadrangle; its total production is estimated by Thomas McDonald as between $7,000,000 and $10,000,000.

The mine has been opened by crosscut tunnels at depths of 340, 395, 420, 520, 640, 725, and 1,080 feet below the outcrop, besides several smaller old workings between the outcrop and the upper level. As two principal veins have been developed to a maximum distance of 1,400 feet, besides drifts on minor veins, and the longest crosscut is about 2,300 feet, it follows that there are several miles of workings. In the short time available it was impossible to make more than a very hasty examination of a comparatively small part of the mine.

There is a 10-stamp mill with two Wilfley tables, capable of handling about 20 tons a shift. In September, 1912, the force consisted of four miners and seven lessees.

The country rock is slate of the Bragdon formation, cut by a large number of irregular intrusions that represent different types, including both diorite and soda granite porphyry.

There is almost everywhere evidence of motion along the slate and porphyry contacts. At only one place was an intrusive contact without gouge seen. The veins lie for the most part in slate but also cut the porphyry. Besides numerous small veins and stringers two principal veins have been worked; these are parallel and have an average strike of N. 80° E. The northern vein, the Monte Cristo, dips steeply to the north; the Last Chance, 200 feet south of the Monte Cristo, dips south at angles between 60° and 80°. The width is as a rule not over 2 feet, more commonly about 6 inches, but stopes have been taken out to a width as great as 22 feet. Two small veins

---

[1] Mineral Resources U. S. for 1909, pt. 1, U. S. Geol. Survey, 1910.
[2] Idem for 1910.

developed to some extent on the lower levels have strikes of N. 20° W. and N. 50° W.

The ore shoots on the Last Chance vein appear to pitch at a rather flat angle to the east. The veins show slipping along both walls and there is always some gouge present, locally as much as 3 inches. The gouge carries fragments of quartz and a little gold. It is very rare that the ore is frozen to the wall.

Quartz is the principal gangue mineral. Much of it has a cloudy blue-gray color and in many places it is banded, owing to inclusions of slate parallel to the walls. Calcite is comparatively rare in the ore and is more commonly seen in the small irregular stringers in the slate and porphyry near the veins. Manganese oxide is noticeable in the surface quartz. The sulphides consist of pyrite, galena, sphalerite, and arsenopyrite. The concentrates carry about $100 in gold to the ton but form a very small percentage of the ore. Pyrite is the most common sulphide and is found both in the vein and impregnating the slate and porphyry walls. Small stringers of pyrite cut the slates near the veins, and the joint planes of the slate and. porphyry are in places heavily pyritized. Galena and sphalerite show no tendency to migrate into the country rock and are found in the vein, their presence generally indicating rich ore. Although they are rare in the ore as a whole, in a few places the vein is almost entirely made up of these two minerals and pyrite. Arsenopyrite is almost exclusively confined to the altered porphyry, where it occurs both in crystals scattered through the rock and in extremely minute crystals along joint planes. Visible gold is common. All the gold seen was in the quartz close to either an included fragment of slate or a patch of galena. Its fineness is about 0.840. The best ore of the present workings runs over $100 a ton and the general run of ore as stoped is between $20 and $50.

DOG CREEK DISTRICT.

The mines of the Dog Creek district are all within the areas of meta-andesite which lie near the heads of Dog Creek and Stacy Creek. On Dog Creek, just east of the quadrangle, is a mass of intrusive alaskite porphyry. Many dikes of soda granite porphyry cut the meta-andesite and the overlying slates of the Bragdon formation.

DELTA MINE (28).

The Delta group of 30 claims (Delta Consolidated Mining Co., owner; S. D. Furber, manager) lies near the head of Dog Creek, about a mile northeast of the Toll House. The recorded production has been $32,000, which does not include the earlier arrastre workings. A narrow-gage railroad was built from the mine to the town of Delta over which a small amount of ore was shipped to the smelter.

The country rock of the veins is meta-andesite, which is cut by numerous dikes of feldspathic dacite porphyry and a few dikes of alaskite. The slates cap the hills above but are not reached by the veins. A short distance east of the quadrangle is a large mass of intrusive alaskite porphyry.

Numerous small veins have been developed for short distances on different claims. These vary from 1 to 2 feet in width and strike between east and N. 70° E. The dips are vertical or steep to the north. Branching is common and the veins tend to taper out within short distances. The workable ore is in irregular shoots and carries between $8 and $10 a ton in gold.

The usual ore is quartz, generally dark blue-gray in color except where oxidized, with a small amount of calcite and scattered specks of barite. The principal sulphide is pyrite, but small amounts of galena and blende are also present in the unoxidized ore. One of the veins contains small patches of arsenopyrite crystals that are said to have an extremely high gold content. In certain of the veins chalcopyrite is prominent. The sulphides, so far as seen, are associated only with the quartz and are never found in the calcite. Pyrite, however, is present in the altered wall rock as well. Usually the other sulphides occur in small lines cutting the quartz or in irregular patches. Gold visible to the naked eye is seen only in the oxidized ore, where it occurs in manganese-stained cavities in the quartz or with streaks of iron oxide.

Recently copper deposits have been discovered on the property. The ore is supposed to be similar to that of the Shasta copper belt in that it is a replacement of an alaskite dike by copper sulphides. Where seen, the ore consisted of specks of pyrite and chalcopyrite in a gangue of grayish quartz and calcite. In several places there are outcrops of a cellular limonitic gossan that carries a small amount of gold. Where prospected by a tunnel 100 feet below this gossan the ore is a much-altered rock, possibly originally alaskite porphyry, carrying pyrite and barite. It assays about $2 to $3 a ton in gold and is supposed to carry copper as well.

COPPER SNAKE PROSPECT.

The Copper Snake prospect (W. B. Glenn, owner), on Stacey Creek about 2 miles southwest of the Toll House, consists of three small prospect tunnels on a somewhat faulted quartz vein. Not enough work has been done to determine the size, though it appears to be of greater width than the average vein of this region. The vein lies between walls of meta-andesite and soda granite porphyry.

The ore is a mottled grayish quartz carrying irregular masses of chalcopyrite and a subordinate amount of pyrite. A small amount of calcite is found near the footwall. Free gold, some of it in valuable pockets, has been found near the surface, particularly in connection with streaks and patches of manganese oxide. Some of the sulphide ore, according to Mr. Glenn, carried 0.6 ounce of gold and 4 ounces of silver to the ton and 15 per cent of copper.

STACEY MINE (29).

The Stacey mine (J. C. Brown, owner) adjoins the Copper Snake on the south. The total production has been about $45,000, entirely from small pockets.

The country rock of the vein is meta-andesite, much altered in the immediate vicinity of the vein. The lower crosscut shows a flat-lying mass of feldspathic soda granite porphyry with irregular boundaries.

The vein is from 1 to 8 feet wide and strikes about N. 15° W. At its northern limit it is cut off by a vertical fault with a strike of N. 30° E. The ground to the northeast is now being explored in the hope of picking up the continuation of the vein.

The ore is chiefly quartz, which is highly manganiferous and accompanied by small amount of manganiferous calcite. The pockets so far found have been in close association with patches or streaks of manganese oxide. No pyrite or prominent iron oxide staining was seen in the ore.

MINERSVILLE DISTRICT.

With the Minersville district is included the country along Trinity River between Trinity Center and Papoose Creek. Numerous irregular folds have brought the meta-andesite to the erosion level reached by Trinity River and its branches, and there are many small patches of meta-andesite whose contacts with the slate have furnished rich pocket deposits. The contacts in the vicinity of Minersville have been particularly productive, and besides supplying the pockets have furnished much gold to the placers of this neighborhood.

FIVE PINES MINE (30).

The Five Pines mine (Five Pines Mining Co., owner; Lester Van Ness, manager) is situated on the northeast side of Van Ness Creek about 1½ miles above its mouth. The property was discovered in 1896 by H. J. Van Ness and has produced a total of $275,000. The equipment consists of a two-stamp mill, but the gold is so coarse that, according to an estimate by Mr. Van Ness, 80 per cent of the total product has been recovered by hand mortar and pan.

The ore lies along a contact of meta-andesite with the overlying slate and sandstone. The meta-andesite occurs as a low dome, the center of which has been cut through by Van Ness Creek. The contact pitches gently to the northeast and southwest and much more steeply to the northwest and southeast, giving the area a roughly elliptical shape with axes about one-half and one-quarter of a mile in length. The nearest intrusive rock is a dike of fine-grained alaskite porphyry which crosses Van Ness Creek a mile southwest of the end of the andesite area.

The dip of the slates and sandstones is most irregular and variable but in general is away from the meta-andesite. There has been motion along the contact of the slates and meta-andesite, as is shown by the crushing of the slates and the gouge along the contact.

The ore is found in a series of pockets which lie along the slate and meta-andesite contact on the northwest and southeast limbs of the anticline, but not at the crest. The mine includes two distinct deposits; the determining feature of each is the presence of a vein of very low grade manganiferous quartz which cuts the slate and underlying meta-andesite. Associated with the vein are many small quartz stringers.

The gold occurs in calcite along the contact of the slate and meta-andesite. Although the calcite, somewhat mixed with quartz, is nearly continuous along the contact and carries everywhere small amounts of gold, the principal returns are obtained from small but very rich pockets in the calcite. The best of these pockets yielded $45,000 in a vertical distance of 44 feet, and other pockets yielding $15,000 and $10,000 have been mined. All these pockets have been found close to the contact of the slate and meta-andesite, but the ore is everywhere much mixed with slate. The low-grade quartz vein and the numerous small stringers of quartz determine the position of the pockets, as all so far found are at or near points where the quartz stringers or vein meet the quartz at the slate and meta-andesite contact. Some ore is found in the meta-andesite, but it is nowhere farther than 4 feet from the slate. More rarely small pockets have been found in the slate where small stringers of calcite cross the slate and sandstones.

The gangue minerals are calcite, quartz, and barite, the first by far the most abundant, particularly in connection with the rich pockets. In a few of the specimens collected it shows a distinct pinkish tinge, presumably due to manganese. Quartz is intimately mixed with the calcite. The surface ore is honeycombed from the solution of the calcite and is in places stained by manganese oxide. Barite was seen in minute tabular crystals in the pink calcite.

Sulphides are rather rare and include only pyrite and arsenopyrite. Both are said to be auriferous, but owing to their low tenor they are not saved. Arsenopyrite is the more common of the two and occurs in minute crystals along shear planes of the black slate close to the calcite, and more rarely in small radial clusters in the calcite itself. Pyrite was seen chiefly in the slate but also in the calcite close to the slate.

Visible gold is not present in the quartz but occurs in the calcite or between the calcite and black slate. In the calcite it appears to be deposited along the cleavage planes and was not seen far from the slate. A hand specimen of ore from this mine on exhibition in the rooms of the State Mining Bureau at San Francisco shows calcite in which the cleavage planes are outlined with gold, the planes being so close together as to give the effect of as great a volume of gold as of calcite in the specimen.

The main adit is 77 feet above the level of Van Ness Creek and 225 feet below the highest outcrop. The contact has been followed down from the adit level for a distance of 225 feet, or to about 43 feet below the stream. A level has been opened at 125 feet down the incline, or about 10 feet below the stream, and pockets have been found in the ground between this and the adit level. No work has yet been done on the lower level, so that it is not certain whether the pockets actually extend below the present water level.

According to Mr. Van Ness all the rich pockets were along water-courses, and the work so far done shows that in deposits of this type gold may be looked for at least as deep as the water level.

MOUNTAIN VIEW PROSPECT (31).

The Mountain View prospect (Fred R. Geddings, owner) consists of several irregular tunnels on the ridge south of Little Bear Gulch. The tunnels are in meta-andesite below the slate; only the upper one follows the contact. Small veins of white quartz, in places deeply stained with manganese oxide, have been followed and stoped to some extent. The largest of these veins is about a foot in width. When crushed and panned the quartz shows free gold in fine colors. The ore has been worked in an arrastre on Little Bear Creek, but the total production is unknown.

FAIRVIEW MINE (32).

The Fairview mine (Fairview Mining Co., owner; W. Waldo, lessee) is on the east bank of Trinity River 2 miles southeast of Minersville post office. The ground was located by Mr. Waldo in 1897 and worked by the company from 1901 to 1908 and by the present lessee since 1910. Mr. Waldo estimates that the total production has been about $200,000. A 40-stamp mill was built during the period of company operation but was never run at its full capacity. The mine is opened by tunnels at four levels, only the lowest of which is now accessible.

The vein as far as worked is entirely in slate and has a strike ranging from west to northwest. The dip is 50°–80° N. The contact of slate and meta-andesite dips about 45° E. It is not known whether the vein pinches out or is faulted at the contact, but from the facts that there has been movement along the contact, as may be seen on the accessible adit level, and that none of the stopes reach the meta-andesite, it is believed that the vein is most probably cut off by a fault. At the third level the vein splits, and in the present workings between the third and fourth levels one branch is about 40 feet to the south. The vein is irregular in outline and its maximum width is about 20 feet.

The ore shoots so far stoped pitch at a low angle (about 20°) to the east and consist of small bands about a foot wide close to the slate walls. The ore as mined runs from $7 to $30 a ton. Occasionally very rich pieces of "specimen ore" are found at the contact of the quartz with the slate. The quartz is much mixed with slate fragments. A few manganese stains were seen. There is a small amount of pyrite present but no other sulphides. The pyrite is said to be practically barren.

DEDRICK DISTRICT.

The country rock of the mines of the Dedrick district near the village of Dedrick, in the northwestern part of the quadrangle, is hornblende schist, which here forms a broad belt, cut by masses of quartz diorite and dikes of porphyritic rocks. The high relief allows much of the mine development by tunnels. The district was formerly a much larger producer than at present, but renewed operations on two of the mines promise to give it a more important place in the near future.

GLOBE (33), BAILEY (34), AND CHLORIDE MINES.

The Globe, Bailey, and Chloride mines (Globe Consolidated Mining Co., owner; C. E. Lamb, manager), are situated on the hill northwest of Dedrick and about 4 miles northwest from the summit of Weaver Bally. Only the Globe (33) is now in operation. The present equipment is a small 10-stamp mill on the east side of the mountain, at an elevation of about 6,500 feet. Power is supplied from the company's plant above Dedrick. The principal difficulty is shortage of water during the summer and heavy snows throughout the winter, which prevent the working of the mine between November and April.

The vein lies in a belt of hornblende schist between two areas of granodiorite. The schist is composed of hornblende with a minor amount of quartz. Near the Globe mine the strike of the schistose banding is N. 55° W. and the dip 60° SW., practically the same as for the whole hill. In a few places small pegmatitic lenses follow the trend of the schistosity. Near the Globe vein a small dike of diorite porphyry cuts the schist and near the Chloride workings are dikes of alaskite porphyry and soda granite porphyry.

The Globe vein is opened by tunnels at elevations of 6,310 and 6,175 feet. The upper tunnel is the longest and has been driven a distance of 1,700 feet, nearly through the hill.

The vein consists of a series of quartz lenses in a zone of much-sheared schist, striking S. 55°–70° W. on an average dip of 60° SE. The intense shearing and consequent drag of the schists make it appear as if the ore lenses were parallel to the schistosity, whereas in reality the strike is nearly at right angles to it. The lenses have an average length of about 200 feet, a maximum of 400 feet, and a width of 8 or 10 feet. The widest portion of the largest lens is 36 feet wide. Parallel lenses in both footwall and hanging wall have been crosscut but not yet developed. The longest interval without quartz between any two lenses is about 50 feet, though it is rare that such an interval exceeds 20 feet. Where the quartz pinches out a well-marked zone of talcose material about a foot wide leads to the next lens. Under the microscope this material is seen to be an aggregate of chlorite, sericitized feldspar, calcite, and a little quartz and pyrite.

The ore is white quartz, much shattered and friable and generally stained with iron and manganese oxides. Another nonmetallic mineral is albite feldspar, which, however, is extremely rare. It occurs only in contact with small fragments of schist included in the quartz, and is much kaolinized. These included fragments of schist are sharp and angular and there is no evidence of replacement. Calcite is present in small amount.

Pyrite is the only sulphide present and is found in irregular patches, generally near the hanging wall. It is auriferous and is saved, though no concentrates have yet been shipped. The gold of this mine is more finely divided than usual and is not in large enough pieces to be visible to the naked eye. About $14 a ton is recovered on the plates. The ore shoots within the lenses are irregular, but the best ore is generally found along the hanging wall. The presence of spots and streaks of manganese oxide is also regarded as a favorable indication.

The Bailey mine (34) lies about a mile to the southeast of the Globe, on the west side of the divide. The upper workings are caved, but a tunnel is being run 60 feet below the lowest of these to crosscut the Globe vein at an elevation of about 5,720 feet. In

September, 1912, this tunnel had a length of 1,100 feet and was expected to reach the Globe ore at about 200 feet farther. At 600 feet from the mouth a 5-foot vein of white quartz carrying a little calcite was crossed. This appears to be the same as that of the upper workings.

The workings of the Chloride mine were entirely caved, but so far as could be seen from the dump and ore bin the ore and country rock are similar to those of the Bailey and Globe.

The company is now erecting a modern stamp mill and cyanide plant north of Dedrick. When the Bailey tunnel reaches the Globe vein it will become the working level of the mine and winter work will then be possible.

CRAIG MINE (35).

The Craig mine (Craig Mining Co., owner; C. E. Lamb, manager) is about 2 miles southeast of Dedrick and a mile west of the border of the quadrangle. It is opened by tunnels 70 feet apart, the longest of which is about 1,100 feet long.

The vein is entirely within the hornblende schist, which here strikes N. 55° W. and dips 70°–85° S. About half a mile to the west is an intrusion of dacite porphyry. The granodiorite lies about a mile to the east.

As in the Globe mine, the vein is a series of quartz lenses and cuts the schist at almost a right angle. The quartz, however, nowhere entirely pinches out, as in the Globe, but varies from 3 inches to 6 feet in width. The gangue consists of quartz with a very subordinate amount of calcite. The quartz is white and dense in the wider parts of the vein; elsewhere it is banded with altered and partly replaced schist and in places is a cloudy gray in color. Calcite was seen only in small patches close to the walls.

Pyrite is the principal sulphide mineral and is much more plentiful than in the Globe. As a rule, it is close to the walls and in the partly replaced schist included in the vein. Much pyrite is scattered through the white quartz, however, some of it in large crystals. These large crystals, some of which are over an inch square, are broken and veined by quartz. A small amount of chalcopyrite is scattered through the quartz with the pyrite.

It is said that the ore averages above $20 a ton but is extremely streaky and irregular. Visible gold is rarely seen.

A 10-stamp mill and cyanide plant are being erected on the property.

SUMMARY.

During late Jurassic or early Cretaceous time the Paleozoic rocks of the Weaverville quadrangle were intruded by masses and dikes of igneous rock. Among the most important of these is the batholith of quartz diorite and granodiorite, which was followed by a series of silicic and basic dikes. The period of fissuring and ore deposition followed the intrusion of the dike rocks, and the veins appear to be genetically connected with certain types of dike rock, particularly the soda granite porphyry.

Native gold is the principal valuable mineral of the veins. The sulphides are auriferous, but their total volume is too small to form an important part of the ores.

The fissure veins in or near the slate of the Bragdon formation are usually associated with dikes or masses of soda granite porphyry. The vein filling is made up of quartz and calcite, with small amounts of galena, sphalerite and arsenopyrite, as well as free gold, which is locally present in flakes large enough to be readily visible. The veins are persistent both in dip and strike and below a shallow zone of surface enrichment show no marked change in character with depth. The rich surface zone is probably due in great part to the solution of the calcite, which has left the quartz correspondingly enriched, and to a less extent to solution and redeposition of the gold.

Comparatively few of the veins in the quartz diorite and alaskite porphyry were studied. In these calcite is less common and pyrite is almost the only metallic mineral present.

The pocket deposits are found almost entirely along faulted contacts of the slate and meta-andesite. It is believed that these deposits are of surficial origin, and that the gold originally present in small quartz veins or in pyrite has been taken into solution by the acid surface waters through the agency of manganese oxide and precipitated by the carbon of the slates, and that the process of pocket formation was facilitated by the neutralization of the descending auriferous waters through the solution of calcite.

www.ingramcontent.com/pod-product-compliance
Lightning Source LLC
Chambersburg PA
CBHW031730210326
41520CB00042B/1746